Math *for* Real Kids

Second Edition

Problems, Applications, and Activities for Grades 5-8

David B. Spangler

A
GOOD
YEAR
BOOK™

Good Year Books
Tucson, Arizona

Good Year Books

Our titles are available for most basic curriculum subjects plus many enrichment areas.
For information on other Good Year Books and to place orders, contact your local
bookseller or educational dealer, or visit our website at www.goodyearbooks.com.
For a complete catalog, please contact:

Good Year Books
PO Box 91858
Tucson, AZ 85752-1858
www.goodyearbooks.com

Cover illustration by Donna Kae Nelson.
All other illustrations by Doug Klauba.
Book design by Dan Miedaner.

Preface

Math for Real Kids is a sourcebook of rich problems, applications, and activities for students at grades 5 through 8. All students at these grades should benefit from using this material because the book provides a wide assortment of the kinds of problems that they are likely to encounter in their everyday lives. Teachers continually rate problem solving as the most important aspect of the mathematics curriculum, so this text should be a valuable resource for them to help their students build confidence in solving problems. Parents also realize that their children need to be better problem solvers—so this book should also be a useful tool for them when they work with their children.

Most problem sets in this book focus on a specific theme, such as scoring in bowling, nutrition, computer memory, camping, energy, sports statistics, consumerism, presidential elections, secret codes, travel in a time machine, making change for a dollar, and more. The themes, many of which are inter-disciplinary, provide for an in-depth treatment of an application. They also provide motivation. Useful formulas are used throughout.

Answers are provided in the back of the book for all problems. Most lessons assume that students have at least some proficiency with the skill being applied. Some lessons include some instructions on the content (such as in the areas of statistics, probability, and spreadsheets) when the content may not be universally known.

Problem information is often presented in charts, tables, maps, or graphs. This parallels real-life situations in that the information required to solve a problem does not always occur in the exact order in which it is needed. Often students are asked to use estimation, make generalizations from patterns, and apply visual thinking.

Math for Real Kids is organized according to the Mathematics Content Standards identified by the National Council of Teachers of Mathematics. These standards include Number and Operations (for whole numbers, decimals, fractions, proportions, percents, and integers), Data Analysis and Probability, Measurement, Geometry, and Algebra. This organization should make it easy for teachers to select lessons—and directly use them—to support their curriculum.

Teachers should allow the use of calculators with this material as they would normally allow their use. If, for the most part, students do the book without calculators, they will gain a tremendous amount of experience doing all sorts of paper-pencil computations in motivational settings. If students primarily do the book with calculators, they will be able to focus on the problem-solving strategies rather than on the computations. This book may be used successfully either way.

Subtitles for the lessons are provided in the Table of Contents to indicate the skills that are being applied. Most lessons include problem sets that involve various operations, including problems that require multiple steps. Even those lessons that essentially apply a specific operation often include problems that require previously learned skills. Thus, students are continually being challenged to develop strategies for determining which operation or operations to use.

A deliberate effort has been made to diversify the problems so that students are exposed to the different types of problems that can be solved with a given operation. For example, there are four types of subtraction problems. Because students often don't realize that these situations exist, it might be useful to outline them here as follows:

- *Take-Away Type:* You know how many there are in all and how many are taken away. You are to find how many are left.

- *Comparison Type:* You know how many are in one group and how many are in another group. You are to find how many more are in one group than are in the other.

- *Additive Type:* You know how many are needed in all and how many you have. You are to find how many more are needed.

- *Separation Type:* You know how many are in a group and how many of a particular kind are in that group. You are to find how many in the group are not of that kind.

Students are exposed to the following types of multiplication situations:

- *Part-Whole Type:* Find the total when you join groups of the same size.

- *Comparison Type:* Find how large something is when it is so many times as large as a given quantity.

- *Area Type:* Find how many objects are in an array when there are so many rows (or columns) of the same size.

Students are exposed to the following types of division situations:

- *Part-Whole Type—How many groups?:* You know how many there are in all and how many are in each group. You are to find how many groups there are of the same size.

- *Part-Whole Type—How many are in each group?:* You know how many there are in all and how many groups of the same size there are. You are to find how many are in each group.

- *Comparison Type:* You know the size of each of two groups. You are to find how many times as large one group is than the other.

To help decide which operation to use to solve a problem, it often helps to think about what *action* could be done with objects to model the problem.

An important element of this book is the use of humor in lesson titles, problem situations, and cartoons. The humor is designed to help students enjoy and feel good about their work in math.

A section of activities is provided at the back of the book. Each activity extends a specific lesson's theme or content. The activities are open-ended, provide high interest, and lead to further exploration. Activities are listed in the Table of Contents under their related lessons.

In closing, the major elements of this book are applications, humor, and activities. The first three letters of those words form the acronym AHA. A goal of this book is for students to conclude: "Aha! This math really is for real kids, it's really important, and I can really do it!"

Acknowledgments

I would like to thank my wife, Bonnie, and my children, Ben, Jamie, and Joey, for all their love, support, and encouragement during the development of this book. I would also like to thank Bobbie Dempsey and Jenny Bevington of Good Year Books and Mark Spears of Scott Foresman for their professional expertise on this project.

Table of Contents

v

Data Analysis

Fractions and Probability

Measurement

Pre-Algebra

Activities

Answer Key

Name: _____   Date: _____

Know Your Bowling Score!

· ·

In bowling, you throw 2 balls in each frame, except when you get a *strike*. A *strike* occurs when you knock down all 10 pins on the first throw.

In frame 1 of the game below, you knocked down 6 pins on the first throw and 3 pins on the second throw, so your score for frame 1 is **9.** In frame 2, you knocked down 8 pins on the first throw and 0 pins (shown by the —) on the second throw. Your score in frame 2 is 9 + 8, or **17.**

In frame 3 you knocked down 5 pins on the first throw and the rest of the pins on the second throw. This is called a *spare* (shown by the /). A *spare* counts for 10 pins, plus the number of pins knocked down with the *first* ball in the next frame. Since you knocked down 4 pins on the first throw in frame 4, your score for frame 3 is 17 + 10 + 4, or **31.**

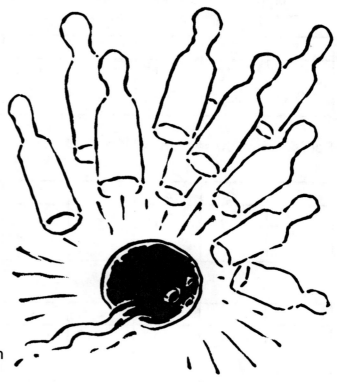

1. Complete the scoring for the game below.

1	2	3	4	5	6	7	8	9	10
6 | 3	8 | –	5 | /	4 | 5	6 | /	9 | –	– | /	2 | 6	7 | /	5 | 4 |
9	17	31							

A *strike* (shown by an X) counts for 10 pins, plus the number of pins you knock down with the next *two* balls. Your score for frame 2 in the game below is 9 + 10 + 7 + 2, or **28.**

2. Complete the scoring for the game below. Note that when you get a spare in the tenth frame, you get one extra throw.

1	2	3	4	5	6	7	8	9	10
9 | –	| X	7 | 2	6 | /	9 | –	| X	5 | 4	– | /	| X	5 | / | 9
9	28								

In frame 2 of the game below, you got a strike. On the next two throws you got another strike (in frame 3) and 4 pins (first throw in frame 4). So your score for frame 2 is 8 + 10 + 10 + 4, or **32.** Your score for frame 3 is 32 + 10 + 4 + 5, or **51.**

3. Complete the scoring for the game below. Note that when you get a strike in the tenth frame, you get two extra throws.

1	2	3	4	5	6	7	8	9	10
8 –	X	X	4 5	6 /	X	X	3 5	1 /	X 3 4
8	32	51							

Score both of these games.

4.

1	2	3	4	5	6	7	8	9	10
2 /	9 –	X	X	7 /	– 9	5 /	X	X X 5 2	

5.

1	2	3	4	5	6	7	8	9	10
X	X	X	6 3	X	– /	5 4	X	X X X X	

6. What is the highest possible final score you can get in a game of bowling? _____

Counting Calories

● ●

	Males			Females	
Age	Weight	Calories	Age	Weight	Calories
7–10	62	2,000	7–10	62	2,000
11–14	99	2,500	11–14	101	2,200
15–18	145	3,000	15–18	120	2,200
19–24	160	2,900	19–24	128	2,200

A *calorie* is a unit for measuring the amount of energy supplied by food. The table at the right shows the daily calorie needs for people of certain ages and weights.

1. A 19-year-old man who does very heavy work needs 1,450 calories a day more than his usual amount. How many calories a day does such a man need?   _____

The table below lists how many calories are provided by various foods. Use the table to find daily calorie intakes on the next page.

Bread, Cereal, Rice, and Pasta Group

Food	Calories
bagel, plain (1)	200
bread, wheat (2 slices)	110
bread, white (2 slices)	126
cake, plain with frosting (1 slice)	445
cornflakes, unsweetened (1.1 oz)	110
cornflakes, sweetened (1.1 oz)	120
doughnut, iced (1)	210
macaroni, hot (1 cup)	190
noodles, cooked (1 cup)	200
pancakes (3 with syrup)	360
pie, apple (1 slice)	377
popcorn, air-popped (1 cup)	30
tortilla, corn (1)	65

Meat, Poultry, Fish, Eggs, & Nuts Group

Food	Calories
bologna (2 slices)	180
chicken, fried (2 drumsticks)	390
egg, fried (1)	110
egg, hard-cooked (1)	75
fish sticks, frozen, reheated (2)	140
hamburger ($\frac{1}{4}$ lb)	334
peanut butter (1 tablespoon)	95
roast beef, lean (2.6 oz)	135
tuna fish, canned in oil (3 oz)	165
turkey, roasted (5 oz)	240

Milk, Yogurt, and Cheese Group

Food	Calories
cream cheese (2 tablespoons)	70
ice cream (1 cup)	270
milk, skim (1 cup)	87
milk, 2% (1 cup)	130
milk, whole (1 cup)	166
milk shake (10 oz)	355
yogurt, fruit filled (8 oz)	230

Fruit and Vegetable Group

Food	Calories
apple (1)	100
banana (1)	104
carrots, small (2)	50
celery (2 stalks)	14
corn, cooked (1 ear)	85
orange juice (1 cup)	110
potato, baked (1)	98
potatoes, fried (20)	394
salad and dressing	119

Fats, Oils, and Sweets

Food	Calories
candy, milk chocolate (1 oz)	145
cola(12 oz)	160
jams and preserves (1 tablespoon)	55
margarine (1 tablespoon)	100

Use the calorie table to find each person's calorie intake for a certain day.

2. *Charlie Calorie, age 12, weight 98 lb*

Food	Calories
Breakfast	
doughnut, iced (1)	_____
pancakes, 3 with syrup	_____
milk, whole (1 cup)	_____
orange juice (1 cup)	_____
Lunch	
hamburger ($\frac{1}{4}$ lb)	_____
potatoes, fried (20)	_____
milk shake (10 oz)	_____
candy, milk chocolate (1 oz)	_____
Dinner	
chicken, fried (2 drumsticks)	_____
salad and dressing	_____
corn, cooked (1 ear)	_____
cola (12 oz)	_____
pie, apple (1 slice)	_____
popcorn, air-popped (1 cup)	_____
TOTAL	_____

3. *Nancy Nutrition, age 13, weight 100 lb*

Food	Calories
Breakfast	
cornflakes, unsweetened (1.1 oz)	_____
egg, fried (1)	_____
milk, 2% (1 cup)	_____
banana (1)	_____
orange juice (1 cup)	_____
Lunch	
bread, white (2 slices)	_____
peanut butter (1 tablespoon)	_____
jam (1 tablespoon)	_____
apple (1)	_____
milk, 2% (1 cup)	_____
Dinner	
turkey, roasted (5 oz)	_____
potato, baked (1)	_____
margarine (1 tablespoon)	_____
corn, cooked (1 ear)	_____
carrots, small (2)	_____
milk, 2% (1 cup)	_____
ice cream (1 cup)	_____
TOTAL	_____

4. Whose calorie intake was more than his or her daily need? _____

Name: _____ Date: _____

Presidential Playing Field

....

The president of the United States is elected by the *electoral college.* States with large populations have more electoral votes than states with small populations.

The candidate who receives the greatest number of popular votes in a state usually gets *all* of the state's electoral votes. To become president, a candidate must receive a *majority* (more than half) of the electoral votes. Since 1964 (when Washington, D.C., was added to the electoral college), the total number of electoral votes has been 538.

Electoral Votes for President
(based on the 2000 Census)

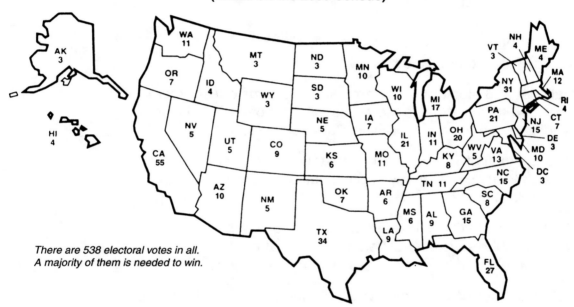

There are 538 electoral votes in all.
A majority of them is needed to win.

1. How many more electoral votes does New York (NY) have than New Jersey (NJ)? _____

2. How many fewer electoral votes does North Carolina (NC) have than Texas (TX)? _____

3. In 1944, California (CA) had only 25 electoral votes. How many electoral votes has it gained since then? _____

4. The six states with the most electoral votes are California, Texas, New York, Florida (FL), Pennsylvania (PA), and Illinois.

    a. How many electoral votes do these six states have in all? _____

    b. How many electoral votes do the rest of the 44 states and Washington, D.C., have in all? _____

Use this table of selected presidential results to answer Questions 5–9.

5. **a.** In 1888, who had more popular votes? _____

    b. How many more? _____

    c. Who had more electoral votes in 1888? _____

    d. How many more? _____

    e. Who won the 1888 election? _____

Selected Presidential Elections

Year	Major Candidates	Popular Votes	Electoral Votes
1888	B. Harrison*	5,444,337	233
	G. Cleveland	5,540,050	168
1992	B. Clinton*	44,909,899	370
	G. Bush	39,104,545	168
	R. Perot	19,742,267	0
2000	George W. Bush*	50,456,169	271
	Al Gore	50,996,116	266
	Ralph Nader	2,831,066	0

(One member of the electoral college chose not to cast a vote for any candidate in 2000.)

*Indicates the winner

6. In which election year were the top two candidates closest in the . . .

    a. number of popular votes received? _____

    b. number of electoral votes received? _____

7. Recall that 538 electoral votes are cast, and a *majority of* them are needed to win. What is the *fewest* number of electoral votes a candidate can have and still win: 268, 269, 270, or 271? _____

8. In 1992, a total of 104,426,659 popular votes were cast for all candidates. How many popular votes were *not* cast for Bill Clinton? _____

9. In 2000, suppose Gore (instead of Bush) had won New Hampshire's 4 electoral votes. Would that have given Gore enough electoral votes to win the election? _____

10. In 2004, George W. Bush received 286 electoral votes. John Kerry received 252 electoral votes. Suppose Kerry (instead of Bush) had won Ohio, but Bush (instead of Kerry) had won Hawaii. Which candidate, if any, would have had the majority of the electoral votes? (Refer to the map on page 5.) _____

    Note: if no candidate gets a majority of the electoral votes, the House of Representatives elects the president from the top three candidates.

From *Math for Real Kids* © 2005 David B. Spangler.

Name: _____   Date: _____

A Trip to 1776

• •

Eli Stu Travel and Bonnie "Bon" Voyage
traveled in a time machine. They stopped
at the year 1776. Upon their arrival, a town
meeting was held to give Eli and Bonnie a
chance to meet local citizens.

1. How many years did Bonnie and Eli
    go back in time?

    _____

2. Abigail Adams asked if American
    women ever gained the right to vote.
    Bonnie told her that American women
    didn't gain the right to vote until 1920.
    How many more years would Abigail
    have to "wait"?

    _____

3. Eli told General George Washington that the U.S. Army now has
    730,000 troops on active duty. Washington then quipped, "That's
    720,000 more than I am able to keep together at any one time!"
    How many troops did Washington have?   _____

4. Eli described modern money, saying that Washington is on the
    $1 bill; Jefferson, the $2 bill; Hamilton, the $10 bill; and Madison,
    the $5,000 bill. When Ben Franklin found out which bill carries
    his picture, he exclaimed, "My picture is worth $4,913 less than
    the others combined!" On what bill is Ben Franklin's picture?   _____

5. Bonnie told Sam Adams that Americans today drink enough tea
    in one year to fill 1,980,000 chests. This is 1,979,658 more chests
    than were dumped during the Boston Tea Party. How many chests
    were dumped that day?   _____

6. Bonnie and Eli knew that both John Adams and Thomas Jefferson
    would die exactly 50 years from July 4, 1776. In what year did both
    men die?   _____

7. Bonnie showed mathematician Benjamin Banneker this problem from *Robinson's Progressive Practical Arithmetic*—a textbook that wouldn't be published for another 97 years:

   A man traveled 6784 miles:
   2324 miles by railroad,
   1570 miles in a stagecoach,
   450 miles on horseback,
   175 miles on foot,
   and the remainder by steamboat;
   how many miles did he travel by steamboat?

   a. What is the answer to the problem? _____
   b. In what year was the textbook published? _____

8. Eli read a poem that was written by Phillis Wheatley in 1772. In it, Ms. Wheatley contrasted the American colonies' demand for independence with her status as a slave. Eli informed those present that slavery wouldn't be abolished until 1865. How many years after the poem was written was slavery abolished? _____

9. On July 4, 1776, Bonnie and Eli watched the 56 delegates from the 13 new states sign the Declaration of Independence. The states with the most delegates were Pennsylvania with 9, Virginia with 7, and Massachusetts and New Jersey each with 5. How many delegates did the other 9 new states have in all? _____

10. Bonnie told Betsy Ross that the Continental Congress would adopt the official U.S. flag—with 13 stars and 13 stripes—on June 14, 1777. Eli said that exactly 100 years later, Congress would declare the first celebration of Flag Day. He mentioned that with each new U.S. state, a star would be added to the flag. He told her that the last time the flag was updated was in 1960 when Hawaii became the 50th state.

    a. On what date was Flag Day first celebrated? _____
    b. How many more stars are currently on the flag than were on the flag in 1777? _____
    c. Bonnie said that her home state, Illinois, became a state in 1819. How many years later did Hawaii become a state? _____

Editor for a Day

· ·

Let's pretend that you are the editor of this math book. As the editor, you have to make sure that all the math is correct. (On this page, the answers to the problems are given in bold type.) You must also correct errors in spelling, grammar, punctuation, and style.

This lesson is loaded with errors! Try to correct as many errrors as you can find. Did you already notice one?

Multiply or divide.

1. 4 x 8 = **32**	**2.** 28 ÷ 4 = **7**	**3.** 0 x 9 = **0**	**4.** 7 x 8 = **54**
5. 36 ÷ 9 = **4**	**6.** 1 x 12 = **1**	**7.** 12 ÷ 3 = **36**	**8.** 8 + 6 = **48**
9. 42 ÷ 7 = **6**	**10.** 90 ÷ 9 = **9**	**11.** 11 ÷ 1 = **11**	**12.** 9 x 9 **81**
13. 10 x 11 = **101**	**14.** 45 ÷ 5 = **9**	**15.** 3 x 3 x 0 = **9**	**15.** 0 ÷ 4 = **0**

Solve each problem

16. Gertrude walks 4 miles each day. How many miles does she walk in12 days.   **48 days**

17. A baseball team gets 3 outs in an inning. How mny outs does a team get in 9 inings?   **27**

18. Mrs Smiths class has $96 to spend on class gifts. if each gift costs $8, How many gifts can the class by?   **13 gifts**

19. A sheet of stamps has 10 rows with 10 stampts in each row. You tear off 2 full rows . How many stamps are left in the sheet?   **8 stamps**

20. Start with the number 72. How many times would You have to subtract 6 to reach 0?   **66**

21. Frank Furter cooked 11 hot dogs. He cut each hot dog into 4 pieces. He needs 48 pieces for a party. How many more hot dogs must he cook?   **4 hot dogs**

Name: _____ Date: _____

A Wall-to-Wall Product

• •

Shown below is the floor of a room that has been tiled.

1. You would like to know how many shaded tiles there are in all.

   a. What *simpler problem* could you first solve to help you find the total number?

   _____

   _____

   _____

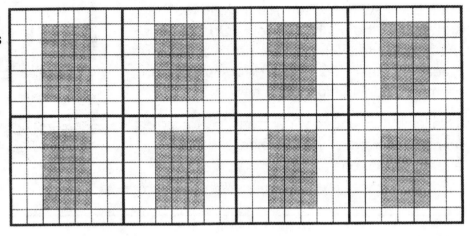

   b. There are 8 sections on the floor outlined by heavy lines. How many shaded tiles are in each of the 8 sections? _____

   c. How many shaded tiles are there in all? _____

2. a. How many tiles are there in all? _____

   b. How many white tiles are there in all? _____

3. a. Suppose each tile costs $5. How much would the tiles cost all together? _____

   b. Suppose each white tile costs $5 and each shaded tile costs $7. How much would the tiles cost all together? _____

4. a. Suppose you would like to tile a room that is twice as long and twice as wide as the above room. How many tiles, like the ones shown, would you need? _____

   b. How many tiles would you need for a room that is 3 times as long and 3 times as wide as the room shown above? _____

   c. A room is 4 times as long and 4 times as wide as the room shown above. How many *times as many tiles* are needed to tile this room than the room shown? _____

From *Math for Real Kids* © 2005 David B. Spangler.

A Bit of Computing

. . . .

Character	Byte
A	01000001
B	01000010
1	00110001
2	00110010

Computers store information in the form of binary digits. These are strings of 0s and 1s. Each binary digit is called a *bit*. A string of

8 binary digits forms a *byte*. A byte is the amount of information needed to store one character from the keyboard. Examples of bytes are shown above.

1. How many bits are in 1,000 bytes? _____

2. A *kilobyte* (1K) is 1,000 bytes. Because of the way chips are made, the *exact* number of bytes in one kilobyte is 2 x 2 x 2 x 2 x 2 x 2 x 2 x 2 x 2 x 2, or _____ bytes.

3. During the early 1980s, many home computers had only 64K of memory. *Exactly* how many bytes is that? _____

4. A *megabyte* (1M) is about 1,000,000 bytes. The *exact* number of bytes in one megabyte is 1,024 x (2 x 2 x 2 x 2 x 2 x 2 x 2 x 2 x 2 x 2), or _____ bytes.

5. Many home computers come with 256M of memory (random-access). *Exactly* how many bytes is that? _____

6. A *gigabyte* (1G) is about 1,000,000,000 bytes. *Exactly* how many bytes equal one gigabyte? _____

7. Pretend you are a computer! Follow the instructions in the program below. Print the results in the "Print-Out" at the right when the program tells you to do so.

PROGRAM	PRINT-OUT
Step	
1. Let *N* = 2.	
2. Multiply the value of *N* by 9.	_____
3. Multiply the result from **Step 2** by 12,345,679.	
4. PRINT the result from **Step 3**.	_____
5. Add 1 to the value of *N*. This will give you a new value of *N*.	
6. If *N* is less than 6, go to **Step 2**.	_____
7. End of program	
	_____

Camp Quo-Tent

• •

Congratulations! You are one of 12 counselors hired at Camp Quo-Tent. Each of you will work all three sessions of this overnight camp. You will be paid $1,176 for working all three sessions.

The cost for campers is $2,100 for a three-week session. The cost is $1,470 for the two-week session.

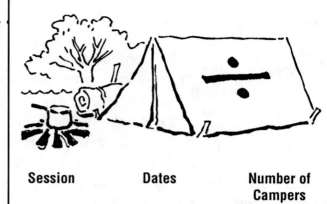

Session	Dates	Number of Campers
A	6/25–7/8 (3 weeks)	156
B	7/9–7/29 (3 weeks)	168
C	7/30–8/11 (2 weeks)	144

1. Counselors will be paid once a week with equal pay checks. How much will you earn each week? _____

2. For each session, the campers will be divided up equally among the counselors. How many campers will there be per counselor during . . .
   a. Session A?_____   **b.** Session B?_____   **c.** Session C?_____

3. **a.** It costs a lot of money to attend this camp. Find the cost per day for a camper during Session B.   _____

   b. Find the cost per day for a camper during Session C.   _____

4. **a.** How many campers will there be in all three sessions combined?   _____

   b. If the campers could be spread out equally among each session, how many campers would there be in each session? (This is the average number per session.)   _____

5. You supervise 41 campers to play baseball. You are to form teams of 9 players each. The rest will be umpires. How many campers will be umpires?   _____

6. The campers and counselors will ride in buses on July 15 to go on an overnight trip. If the capacity load on each bus is 48 people, how many buses will be needed?   _____

7. 60 of the campers will sleep in 13 tents on the overnight trip. (The rest will sleep in cabins.) Approximately the same number of people will occupy each tent.

   How many tents will have 4 people each?   _____

   How many tents will have 5 people each?   _____

8. During an activity, you have exactly enough campers to form 8 teams of 15 each. How many teams of 12 each could you form among those campers? _____

9. During her 2 weeks in camp, Rita Letter received 5 letters by E-mail, 3 letters by fax, and 6 letters by regular mail. How many letters did she receive on average per day? _____

Before going scuba diving, you asked your campers to study the chart at the right. It shows the approximate pressure caused by the weight of the water on the diver.

Pressure from Sea Water When Scuba Diving	
Depth	**Pressure on Diver**
sea level	15 lb per square inch
-33 feet	30 lb per square inch
-66 feet	45 lb per square inch
-99 feet	60 lb per square inch

10. **a.** How many times as great is the pressure at -66 feet than it is at sea level? _____

 b. How many times as great is the pressure at -99 feet than it is at sea level? _____

11. "So far in camp we have already done 561 division problems!" said an excited Ray Show. "This is 17 times as many division problems as we did all last year—and 51 times as many as we did 2 years ago!" How many division problems did they do . . .

 a. last year? _____ **b.** 2 years ago? _____

Stuff You Auto Know About Mileage

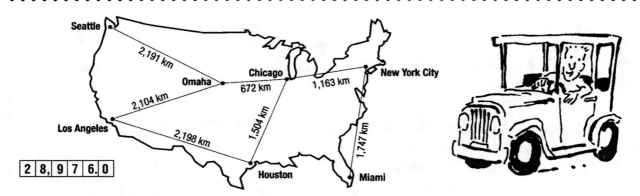

1. **a.** Shown above is the odometer reading in kilometers on
   Mrs. O'Dometer's car after a trip from Los Angeles to Houston.
   What was the reading *before* the trip?   _____

   b. If she used 157 liters of gasoline during the trip, how many
   kilometers did she travel per liter of gasoline?   _____

2. Anita Map traveled from Omaha to Houston by way of Chicago.
   How many kilometers did she travel?   _____

3. Seymour Ocean made 3 round trips between New York City
   and Miami. How many kilometers did he travel?   _____

4. The van Tripp family plans to drive from Seattle to Omaha to
   Chicago to New York City. If they share the driving equally among
   the father, the mother, and the oldest child, how many kilometers
   will each drive?   _____

5. **a.** Mr. Oasis drove from Houston to Chicago. Along the way he
   stopped every 115 kilometers to rest. How many times did
   he stop for rest before reaching Chicago?   _____

   b. How many kilometers did he have left to drive after his last
   stop for rest?   _____

6. Les Fuel and G. I. Guzzle each drove from Chicago to Omaha.
   Les got 12 km per liter with his car, but G. I. only got 4 km per liter
   with his gas guzzler. How many more liters of gas did G. I. use for
   the trip than Les?   _____

7. In O-Bay County, the fine for speeding is $85, plus $4 for each
   mile per hour over the speed limit. How much will a speeder have
   to pay if he or she is caught driving at a speed of 43 mi/h
   in a 25-mi/h zone?   _____

Name: _____   Date: _____

Formula-Driven

• •

A *formula* is a rule or principle written as a mathematical sentence.

This formula shows how the speed of an ant can help you determine the outside temperature.

This formula shows how a cricket can help you determine the outside temperature.

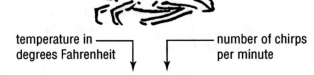

temperature in degrees Fahrenheit   speed in inches per second

$T = (11 \times s) + 37$

temperature in degrees Fahrenheit   number of chirps per minute

$T = (c \div 4) + 37$

1. What is the temperature outside if an ant is traveling at a speed of
   a. 1 in. per sec? _____
   b. 3 in. per sec? _____
   c. 4 in. per sec? _____
   d. 6 in. per sec? _____

2. According to the formula, at what temperature will an ant stop traveling (move 0 in. per sec)? _____

3. Suppose the outside temperature is 59°F. How fast will an ant travel? _____

4. What is the outside temperature if a cricket makes
   a. 204 chirps per min? _____
   b. 196 chirps per min? _____
   c. 172 chirps per min? _____
   d. 140 chirps per min? _____

5. According to the formula, will a cricket chirp faster or slower as the temperature goes up? _____

6. Suppose the outside temperature is 77°F. How many chirps will a cricket make per minute? _____

7. According to the formula, at what temperature will a cricket stop chirping? _____

8. Suppose a cricket chirps 220 times per minute. How fast would an ant be traveling? _____

From *Math for Real Kids* © 2005 David B. Spangler.

9. Sometimes it is fun to use a formula that is made up. This formula tells how many new loads of laundry, *N,* still need to be done in a week:

$$N = F \times L$$

L stands for the number of loads that have already been done during the week.

F stands for the frustration factor.

F = 2 x *c* + *a,* where *c* = the number of children in the family; *a* = the number of adults in the family.

Find how many new loads of laundry still need to be done in a week for a family that has already done 7 loads and has 3 children and 2 adults.

_____

(Based on the formula, the more loads of laundry you do, the more loads you still have left to do!)

Can You Make the Change?

· ·

Here is how you can make change for a customer in a store:

    a. Begin with the *amount due*.

    b. Add amounts until reaching the *amount given to you* by the customer.

    c. Use the fewest possible pennies, nickels, dimes, quarters, and bills.

1. Suppose the *amount due* is $0.32, and the *amount given to you* is $1. In making change, you give the customer **3** pennies and say, "**$0.35**." Then you give the customer **1** nickel and say, "**$0.40**."

   Then you give the customer _____ dime and say, "_____."
   Finally, you give the customer _____ quarters and say, "_____."
   The amount of change is $0.03 + $0.05 + $0.10 + $0.50, or $_____.

Complete the following:

2.

Amount due: **$1.18**	
Amount given to you: **$2**	
Change	**What you say**
_____ pennies	_____
_____ nickel(s)	_____
_____ dime(s)	_____
_____ quarter(s)	_____
The amount of change is **$** _____	

	Amount Due	Amount Given to You	Change						Amount of Change
			$0.01	$0.05	$0.10	$0.25	$1.00	$5.00	
3.	$2.37	$5.00	3	0	1	2	2	0	
4.	$0.84	$1.00							
5.	$0.29	$1.00							
6.	$2.33	$3.00							
7.	$1.78	$5.00							
8.	$4.02	$10.00							
9.	$5.41	$10.00							
10.	$14.58	$20.03							

Name: _____ Date: _____

To Life!

· ·

Years of Life Expected at Birth in the U.S. (All races combined)		
Year	Males	Females
1920	53.6	54.6
1930	58.1	61.6
1940	60.8	65.2
1950	65.6	71.1
1960	66.6	73.1
1970	67.1	74.7
1980	70.0	77.5
1990	71.8	78.8
2000	74.4	79.8

This table tells how many years a person born in a particular year can expect to live. This is a person's *life expectancy.*

1. During which year were males and females . . .
   a. closest in life expectancy? _____
   b. furthest apart in life expectancy? _____

2. The *range* for a set of numbers is the difference between the largest number and smallest number in the set.
   Find the range for the . . .
   a. male life expectancies. _____
   b. female life expectancies. _____

3. **a.** Between which two years in the table did the life expectancy for males increase the most? How much was that increase? _____ ; _____
   b. Answer the same questions for the females. _____ ; _____

4. How much more did the life expectancy increase for males between 1990 and 2000 than it did for females? _____

5. Write a couple of sentences describing any trends you notice in the table.

   _____

   _____

   _____

6. When comedian George Burns was born in 1896, his life expectancy was 42.5 years. When he celebrated his 100th birthday (on 1/20/96), how many years beyond his life expectancy had he lived? _____

7. A person who smokes cigarettes should subtract 6.16 years from his or her life expectancy. Find the life expectancy for a male smoker and for a female smoker born in 1980. _____ ; _____

8. Use the information in the table on the previous page to estimate *your* life expectancy. _____

In the table below, the data for males and females are combined.

Years of Life Expected at Birth			
Country	**Years**	**Country**	**Years**
Canada	79.4	Italy	79.0
China	71.4	Japan	80.7
Egypt	63.3	Mexico	71.5
India	62.5	Rwanda	39.3
Israel	78.6	U.S.	77.1

9. How much would the life expectancy in Rwanda have to increase to equal that in Egypt? _____

10. **a.** In the space below list the countries with their life expectancies in order, from lowest to highest.

   b. Find the *range* for the set of data. _____

Take Your Cue from Coupons

· ·

When you go grocery shopping, you can save money using coupons. There are two types of coupons. *Manufacturers'* coupons are issued by makers of products. *Store* coupons are issued by the store where you are shopping.

Use the above coupons with Questions 1–5. The prices for the items are given at right.

Note: If you have coupons from both the manufacturer and the store for the same product, you may use *both* of them at the same time.

Shell's peanut butter, 18 oz	$2.39
generic peanut butter, 18 oz	$1.77
Admiral Soggy's oat cereal, 12 oz	$4.42
generic oat cereal, 12 oz	$2.89
D-Licious vanilla wafers, 12 oz	$2.51
generic vanilla wafers, 12 oz	$1.69
Worm's applesauce, 24 oz	$2.74
generic applesauce, 24 oz	$2.25

1. **a.** Which costs more: a jar of *Shell's* peanut butter purchased with coupons or a jar of generic peanut butter? _____

   b. How much more? _____

From *Math for Real Kids* © 2005 David B. Spangler.

2. **a.** How much cheaper is it to buy *Worm's* applesauce with a coupon than it is to buy generic applesauce?

   b. Suppose you do not have a coupon. How much cheaper would it be to buy the generic applesauce? _____

3. **a.** Find the cost of 2 boxes of generic oat cereal. _____

   b. Find the cost of 2 boxes of *Admiral Soggy's* oat cereal with coupons. _____

4. You want 1 jar of peanut butter, 1 box of vanilla wafers, and 1 jar of applesauce. How much would these cost if you bought the . . .

   a. brand-name products *without* coupons? _____

   b. brand-name products *with* coupons? _____

   c. generic products? _____

5. If you are dissatisfied with a jar of *Worm's* applesauce, they will give you double your money back (what you paid). Suppose you buy a jar with a coupon and are dissatisfied. How much money will you get back? _____

Checking Your Checking Account

. . . .

Each month your bank sends a *bank statement.*
The bank statement should be compared with your
checkbook register to make sure that you and the
bank agree. A bank statement for Ima D. Positor is
shown on the next page. Ima's checkbook register
is shown below it.

Use the bank statement and checkbook
register on the next page to complete the
questions below.

1. Balance the checkbook by doing the following:
    a. Write a ✔ in the checkbook for each check and deposit that
       appears in the bank statement.
    b. Find the sum of deposits listed in the checkbook that are *not*
       listed in bank statement.   _____
    c. Find the sum of the checks listed in the checkbook that are *not*
       listed in bank statement.   _____
    d. Add **c** to the last balance shown in the checkbook.   _____
    e. Subtract **b** from **d**.   _____
    f. Compare your result in **e** to the Ending Balance in the bank
       statement. Do they agree?   _____
       If not, what must be done to balance the account? Do it.

       _____

       _____

    g. What should be the correct final balance in Ima's checkbook?   _____

2. Pretend you are Ima D. Positor. Suppose today is April 5, and you
    are about to buy some fruit for $29.78 at Produce with A-peel.
    You plan to use check #645 from your checkbook for this
    transaction. Record the transaction in the check register on the
    next page. What is the new balance in your account?   _____

From *Math for Real Kids* © 2005 David B. Spangler.

Ima D. Positor's bank statement covers the month of March. Ima's checkbook has entries past March because transactions have taken place since the bank statement came out.

Bank Statement for Ima D. Positor

Bank Statement			GotchaMoney $tate Bank
Beginning Balance 947.52	**Total Amount of Checks and Fees** 328.22	**Total Amount of Deposits** 542.89	**Ending Balance** 1162.19

Checks and other Debits		Deposits	Date	Balance
		100.00	3/3	1047.52
Check #639	157.88		3/7	889.64
Check #640	10.75		3/14	878.89
		68.26	3/25	947.15
Check #642	151.09		3/30	796.06
		374.63	3/30	1170.69
Monthly Checking Fee	8.50		3/31	1162.19

Checkbook Register for Ima D. Positor

	Check Number	Date	Transaction	Amount of Payment	✔	Amount of Deposit	Balance 947.52
C		3/2	Deposit			100. 00	1047. 52
H	639	3/5	Chuck Roast	157. 88			889. 64
E	640	3/11	Dina Soar	10. 75			878. 89
C	641	3/17	Josey Wosey, Inc.	38. 42			840. 47
K		3/23	Deposit			68. 26	908. 73
B	642	3/28	Dr. I. Stitch	151. 09			757. 64
O	643	3/30	Plug Outlet Store	12. 68			744. 96
O		3/30	Deposit			374. 63	1119. 59
K		4/1	Deposit			100. 00	1219. 59
	644	4/2	Kent Reed	25. 05			1194. 54

More Power to You

• •

This table shows the number of calories a person burns during various activities.

Activity	Calories burned per hour for each kilogram of mass
sleeping	0.9
reading	1.8
walking	2.4
bicycling	3.9
dancing	5.7
jogging	10.1

1. Suppose your mass is 40 kilograms. In 1.75 hours of bicycling, you will expend 1.75 x 3.9 x 40, or _____ calories.

2. How many calories will a 40-kilogram person burn . . .
   a. during 8 hours of sleep?   _____
   b. during 1.5 hours of dancing?   _____

3. How many calories will a 42.5-kilogram person burn . . .
   a. during 2.5 hours of reading?   _____
   b. during 0.4 hour of jogging?   _____
   c. during 1.3 hours of walking?   _____

4. Suppose your mass is 49.5 kilograms and you eat a 375-calorie sundae. Will you burn the 375 calories if you go bicycling for 2 hours?   _____

The *watt* is a unit used to measure electric power. If you know how many watts an appliance uses, you can compute how much it costs to run it, as described below.

- A 150-watt light bulb left on for 7 hours will use 7 x 150, or 1,050, *watt-hours* of electricity.
- Your electric meter measures electricity in *kilowatt-hours* (kWh). To find the number of kilowatt-hours, divide the number of watt-hours by 1,000. (1,050 ÷ 1,000 = 1.050)
- The cost of electricity is found by multiplying the rate for electricity in your area by the number of kilowatt-hours used. If the rate is 7.9¢ per kilowatt-hour, then running a 150-watt light bulb for 7 hours would cost $0.079 x 1.050 = $0.08295, or about $0.08.

5. Complete this table for appliances in a home where electricity costs 8.2¢ per kWh.

Appliance	Hours Turned On	Watt-hours	Kilowatt-hours (kWh)	Total Cost (nearest cent)
a. 100-watt light bulb	24.00			
b. 6,700-watt electric oven	1.50			
c. 850-watt microwave oven	0.15			
d. 225-watt TV	8.50			

Name: _____ Date: _____

At Any Rate...

• •

A *rate* is the quotient of two quantities of different units of measure. In situations like the following, rates can be calculated:

A. price per can: $3.54 for 6 cans of cola	**B. airplane speed:** 1,540 miles in 2.75 hours	**C. U.S. population density:** 296,013,000 people on 3,536,341 square miles

The rate for one unit of a given quantity is called the *unit rate.* In **A** above, the price per can of cola is

$$\frac{\text{price}}{\text{cans}} = \frac{\$3.54}{6 \text{ cans}} = \frac{\$.059}{1 \text{ can}}, \text{ or } \$0.59 \text{ per can.}$$

A unit rate, such as the price per can or per pound, is also called the *unit price.*

1. **a.** In **B** above, what is the speed in miles per hour?   _____

    b. A car travels 181.8 miles in 3.5 hours. What is its average speed in miles per hour? Round to the nearest tenth.   _____

    c. You go 2.22 miles on a treadmill in 35.2 minutes. What is your speed in miles per minute to the nearest hundredth?   _____

2. Use **C** above to find the U.S. population density in people per square mile. Round to the nearest person.   _____

3. During the 2005 Super Bowl, the cost for a 30-second TV commercial was $2,400,000. Find the cost per second to the nearest thousand dollars.   _____

4. **a.** *Flaky Flakes* cereal comes in three sizes: 10 ounces for $2.45; 15 ounces for $3.47; and 22 ounces for $4.72. Find the unit price for each to the nearest tenth of a cent.

    10 oz _____ 15 oz _____ 22 oz _____

    b. Suppose you have a 75¢-off coupon that is good on any size box. Find the unit price for each again. Circle the unit price that is the *best buy.*

    10 oz _____ 15 oz _____ 22 oz _____

5. Circle the rate of pay below that would give you the highest yearly earnings. Cross out the one that would give you the lowest yearly earnings. Assume that you work 40 hours per week.

    $25,000 per year $2,200 per month $500 per week $13 per hour

Super Sports Statistics I: Baseball and Basketball

· ·

To find a baseball player's *batting average*, divide the number of hits by the number of at-bats. Round to the nearest thousandth. (Omit the zero before the decimal point.)

1. Find the batting average for each player using the data in the table.

Player	At bats	Hits	Average
Ichiro Suzuki	704	262	_____
Manny Ramirez	568	175	_____
Albert Pujols	593	196	_____
Todd Helton	547	190	_____

To find a baseball pitcher's *earned run average* (average number of earned runs allowed in a game), multiply the number of earned runs by 9. Divide that product by the number of innings pitched. Round to the nearest hundredth.

2. Find the earned run average for each pitcher using the data in the table.

Pitcher	Innings pitched	Earned runs	Earned run avg.
Ben Sheets	237.00	71	_____
Carlos Zambrano	209.67	64	_____
Curt Schilling	226.67	82	_____
Johan Santana	228.00	66	_____

To find a team's *win percentage,* divide the number of games won by the total number of games played. Round to the nearest thousandth, omitting the zero before the decimal point. (A team with a .500 win percentage has won as many games as it has lost.)

3. Find the win percentage for each team listed in this table.

2004 Final Results American League East Division			
Team	Games won	Games lost	Win percentage
New York	101	61	_____
Boston	98	64	_____
Baltimore	78	84	_____
Tampa Bay	70	91	_____

Use the formula below to see if a baseball team is winning as many games as it "should be winning." The formula, called the *estimated win percentage,* considers the runs a team scores and the runs it gives up.

$$\textit{Estimated Win Percentage} = \frac{\text{Runs squared}}{\text{Runs squared} + \text{Opponents' Runs squared}}$$

4. In 2004, the New York Yankees scored 897 runs. Their opponents scored 808 runs.

   a. Find New York's estimated win percentage to the nearest thousandth. Hint: "Runs squared" = $(897)^2$, or 897 x 897. _____

   b. In 2004, did New York do better or worse than what the estimate suggests? (Refer to New York's win percentage in Question 3 on the previous page.) _____

You can estimate a basketball team's win percentage by looking at the points it scores and the points it gives up.

$$\textit{Estimated Win Percentage} = \frac{\text{Points}}{\text{Points} + \text{Opponent's Points}}$$

5. During the 2003–04 NBA season, the Detroit Pistons won 54 games and lost 28 games. They scored 7,388 points and gave up 6,909 points.

   a. Use the points data to find the Pistons' *estimated win percentage.* _____

   b. Find the actual win percentage for the Pistons. (Divide the number of games won by the total number of games played.) _____

   c. Did the Pistons do better or worse than what the estimate suggests? _____

A basketball statistic that some people use to rate individual players is called the *Performance Totals (PT).*

PT = Points + Rebounds + Assists + Blocks + Steals – Turnovers – Field Goals Missed – Free Throws Missed

6. Find the 2003–04 *PT* for each of these players. In the table below, "PTS" stands for *points,* "REB" stands for *rebounds,* and so on.

Player	PTS	REB	AST	BLK	STL	TO	FGM	FTM	PT
Tracy McGrady	1,878	402	370	42	93	179	913	102	
Predrag Stojakovic	1,964	508	173	14	108	153	721	31	
Kevin Garnett	1,987	1,139	409	178	120	212	807	97	

Super Sports Statistics II: Football

. . . .

1. You are about to compute a football statistic that isn't easy to *tackle!* It is the *NFL Passer Rating for* quarterbacks in the National Football League. You will use it to rate the three quarterbacks listed in this table (2003 regular season).

Quarterback	Attempts	Completions	Yards	Touchdowns	Interceptions
a. Steve McNair	400	250	3,215	24	7
b. Peyton Manning	566	379	4,267	29	10
c. Daunte Culpepper	454	295	3,479	25	11

Follow the steps below to find the *NFL Passer Rating* for each of the quarterbacks above. Write the result you obtain for each step in the table that follows.

Finding the NFL Passer Rating:

Step 1: Divide the Completions by the Attempts. Subtract 0.3. Then divide by 0.2. Round to the nearest thousandth.

Step 2: Divide the Yards by the Attempts. Subtract 3. Then divide by 4. Round to the nearest thousandth.

Step 3: Divide the Touchdowns by the Attempts. Divide by 0.05. Round to the nearest thousandth.

Step 4: Divide the Interceptions by the Attempts. Subtract this quotient from 0.095. Then divide by 0.04. Round to the nearest thousandth.

Note: The greatest result that is allowed in any step is 2.375. The smallest result allowed is 0. (This keeps any one area from having too much of an influence on the rating.)

Step 5: Add the results from Steps 1–4. Divide by 0.06. Round to the nearest tenth. You have found the NFL Passer Rating.

Quarterback	Result from Step 1	Result from Step 2	Result from Step 3	Result from Step 4	NFL Passer Rating
a. Steve McNair					
b. Peyton Manning					
c. Daunte Culpepper					

From *Math for Real Kids* © 2005 David B. Spangler.

2. Here's how you can make a Super Sports Card! Collect data from a sport or recreational activity in which you participate. If no such data are available, pick a favorite athlete for whom you can obtain such data from a newspaper or almanac. Use the data to compute some statistics. Then use the form below to make your own trading card for you or the athlete. On the front, draw a picture of the person (or attach a 5" x 7" picture).

Super Sports Card

Name _____

Nickname _____

Age _____ **Height** _____

Team _____ **League** _____

Season _____ **Team Record** _____

Coach(es) _____

Position(s) _____

Statistics

Additional Information; Season Highlights

Autograph _____

Name: _____   Date: _____

Wheels and Deals

· ·

1	4,	9	8	8.	7

Before

1	5,	3	3	0.	6

After

1. **a.** The above odometer readings, in miles, were recorded before
    and after a trip. How many miles long was the trip?   _____

    b. 12.9 gallons of gas were used during the trip. Find the average
    number of miles driven per gallon. Round to the nearest tenth.   _____

2. A car with a 16-gallon fuel tank gets 28.8 miles per gallon.
    Its fuel gauge appears at the right. How far can the car go
    with the gasoline that is in the tank right now?

3. Suppose you drive 12,300 miles a year, pay an average price of
    $2.49 per gallon for gasoline, and get 20.5 miles per gallon.

    a. How much would you spend in 1 year for gasoline?   _____

    b. Suppose your car averages 30.75 miles per gallon instead of
    20.5 miles per gallon. How much less would you pay in 1 year?   _____

4. A safe distance to drive behind the car in front of you is 1 car length
    for every 10 miles an hour you are traveling on dry pavement.
    Suppose your car is 5 meters long. At least how many meters
    must you stay behind the car in front of you if you are traveling at . . .

    a. 35 miles per hour _____   **b.** 55 miles per hour _____

5. The formulas below are sometimes used to estimate the dealer's cost *d* for various kinds
    of new cars, when given the sticker price *s*. Find the dealer's cost for the cars with the
    sticker prices given below. Also, for each car find the dealer's mark-up (the difference
    between the sticker price and the dealer's cost). Round each result to the nearest dollar.

    $$d = \frac{s}{1.12} \qquad d = \frac{s}{1.18} \qquad d = \frac{s}{1.25}$$

    Compact car   **Medium-sized car**   **Full-sized car**

    a. a $27,999 full-sized car   _____ ; _____
    b. a $19,750 medium-sized car   _____ ; _____
    c. a $16,660 compact car   _____ ; _____

From *Math for Real Kids* © 2005 David B. Spangler.

Name: _____ Date: _____

Learning about the Mean

• •

The *mean, median,* and *mode* are three kinds of averages that can be used to describe a set of data. They are known as measures of central tendency.

Calculation of the mean: Find the sum of the scores. Divide the sum by the number of scores.

Example:
Find the mean for this set of numbers:
4, 0, 7, 10, 9

The mean is $= \dfrac{4 + 0 + 7 + 10 + 9}{5} = \dfrac{30}{5} = 6$.

In Problems 1–3, find the mean. Round to the nearest tenth, where necessary.

1.	2.	3.	4. Multiple Choice
82	125,000	0.990	Consider this set of scores:
95	150,000	5.750	20, 35, 45, 50, 60
87	200,000	42.015	The mean for this set is . . .
0	4,000,000	82.200	(a) 20.
85	650,000	106.550	(b) a number from 30 to 39.
99	350,000	39.000	(c) a number from 40 to 49.
98		14.707	(d) 50.
		99.880	(e) a number larger than 50.
Mean: _____	Mean: _____	Mean: _____	

5. Consider this set of scores: 10, 20, 30, 40, 50, 60, and 70.
   True or False: The sum of the above scores will not change
   if each is replaced with the number 40.   _____

6. Suppose you know the mean for a set of scores, and you know how many scores are in the set. Do you have enough information to find the sum of the scores? Explain.

_____

_____

Learning About the Median

. . . .

The *median* separates a distribution in half—with half the scores above the median and half the scores below it.

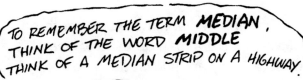

Calculation of the median: Arrange the scores in ascending or descending order.

• If the number of scores is odd: The *median* is the middle score.

• If the number of scores is even: The *median* is the mean of the two middle scores.

Examples:

• Find the median for these scores:
39, 12, 75, 39, 100, 89, 42
Arrange the scores in order.
12, 39, 39, 42, 75, 89, 100
The median is 42.

• Find the median for these scores:
33, 76, 19, 22, 8, 1,000
Arrange the scores in order.
8, 19, 22, 33, 76, 1,000

The median is $= \dfrac{22 + 33}{2} = \dfrac{55}{2} = 27.5$.

In Problems 1–3, find the median (to the nearest tenth, where necessary).

1.	2.	3.	4. Multiple Choice
82	125,000	0.990	There are five scores in a set. The mean is 28, and the median is 19. No two scores in the set are the same. Which is true if the highest and lowest scores are removed?
95	150,000	5.750	
87	200,000	42.015	
0	4,000,000	82.200	
82	650,000	106.550	
99	350,000	39.000	(a) The mean will still be 28.
98		14.707	(b) The median will still be 19.
		99.880	(c) Both of the above,
			(d) There is not enough information to tell what will be true.
Median: _____	Median: _____	Median: _____	

From *Math for Real Kids* © 2005 David B. Spangler.

Name: _____ Date: _____

Learning About the Mode

· ·

Calculation of the mode: The *mode* is the score or item that occurs most frequently in a set. If more than one item occurs with the greatest frequency, each of those items is the mode. If no item occurs with the greatest frequency, the mode *does not exist.*

Examples:

A, B, B, B, C, C, D, B, A, A, C, F	Mode = B
2, 3, 3, 3, 5, 5, 6, 6, 6, 7, 7, 8, 8	Modes = 3 and 6
2, 3, 3, 3, 5, 5, 6, 6, 6, 7, 7, 7, 8, 8	Modes = 3, 6, and 7
2, 4, 6, 9	The mode does not exist.
2, 2, 2, 2	The mode does not exist.
2, 2, 3, 3, 5, 5	The mode does not exist.

Note that in the last three cases above, no score occurs with the greatest frequency.

To help you remember the term mode, think of the word "most." The mode is the item that occurs most often. The word *mode* comes from the French, meaning, "in fashion."

The mode can be determined from a *frequency table.*
A frequency table tells how often each item occurs.

In the table, the number of times each item occurs in a set appears in the frequency column *(f)*. To determine the mode, find the greatest frequency in the column. In the example at the right, the mode is *B* because more scores are *B* than any other score. (Note: This is the same distribution of letters that is given in the first example.)

Grade	f
A	3
B	4
C	3
D	1
F	1

Find the mode or modes.

1.	2.	3.
85, 90, 85, 42, 90, 77, 77, 85, 42, 85, 42, 90 Mode(s) _____	90, 82, 76.5, 82, 90, 76.5 Mode(s) _____	77, 91, 85, 18, 90, 39, 77, 18, 91, 39, 77, 91 Mode(s) _____

4.	Color	f
	Blue	7
	Hazel	3
	Brown	7
	Green	3
	Mode(s): _____	

5. **Multiple Choice.** Circle all that are true,

(a) The median is always one of the scores in a set.

(b) The mean can never be larger than the largest number in the set.

(c) The median is always smaller than the mean,

(d) The mode, when it exists, is always one of the scores in a set.

Name: _____   Date: _____

Finding Averages from a Frequency Table

• •

To reduce pollution, gas usage, and the
amount of traffic, people should carpool when
possible. The results below show how many
people arrived in each car one working day at
the SmogGuzzler Natural Resource Company.

1	1	3	1	4	5	1	5	2	3	7	1	3	2	1	1	6
2	3	4	1	5	3	7	2	5	6	1	2	1	1	1	3	4
1	6	1	1	1	2	1	1	3	1	1	1	4	1	2	1	4

To help see the "big picture," you can organize the data into a frequency table.
A *frequency table* lists each possible score and the frequency with which it occurs.

1. Complete the frequency table. Use tally marks to help you find the frequencies.

Number of people in each car	Tally	Frequency
1		
2		
3		
4		
5		
6		
7		

2. What is the mode?   _____

3. a. How many scores (cars surveyed) are there in all?   _____
    b. The median is the 26th score. Find the median.   _____

4. To find the mean, do the following:

    (a) Multiply each score by its frequency, *f.*

    (b) Add the products.

    (c) Divide the sum of the products by the sum of
       the frequencies. Round to the nearest tenth.
       (You are dividing the number of people by the
       number of cars.)

    The mean is _____.

Score	f	Score x Frequency
1	23	
2	7	
3	7	
4	5	
5	4	
6	3	
7	2	

Problems That Are Just About Average

· ·

1. Suppose you received these scores on math tests: 94 86 92 12 100 99 91 99 83

    Find the mean, the median, and the mode of your scores.

    Mean _____ Median _____ Mode _____

2. **a.** Which of the above measures of central tendency seems to best describe the above
    distribution of math scores: the mean, the median, or the mode? _____
    b. Explain your answer.

    _____

3. Your teacher decides to drop your lowest score (12). Find the mean, median, and mode
    for the remaining 8 scores above.

    Mean _____ Median _____ Mode _____

4. The yearly salaries at a small company are shown below.

$19,000	$20,000	$32,000	$40,000
$20,000	$20,000	$38,000	$46,000
$20,000	$24,000	$40,000	$281,000

    Find the mean, median, and mode salaries.

    Mean _____ Median _____ Mode _____

5. You are representing the employees above at the
    bargaining table. To show that the employees are
    underpaid, which salary would you use: the mean,
    the median, or the mode? _____

6. Discard the $281,000 from the salaries above.
    Find the mean, median, and mode for the
    remaining 11 salaries.

    Mean _____

    Median _____

    Mode _____

7. Study the problems on this page. Which seems
    to be affected most by either a very high score
    or a very low score: the mean, the median, or
    the mode? _____

8. A manager of a shoe store would like to know which shoe size sells best. Which shoe size should she find: the mean, the median, or the mode? _____

9. In a gymnastics meet, four judges score each gymnast. They throw out the highest and lowest scores and find the mean of the two middle scores. This gives the final score. The final score is really the (mean, median, or mode) of the four original scores. _____

10. You take 6 tests, and your mean score is 92.5. How many total points did you earn? _____

11. **a.** Suppose you take 4 tests and earn scores of 81, 83, 88, and 89. What is the minimum score you must earn on the next test to produce a mean of at least 88? _____

   b. Explain how you found the result in 11a.

   _____

   _____

12. Find the mean, median, and mode for the distribution of test scores at the right:

   Mean _____ Median _____ Mode _____

Score	f
100	3
95	1
90	3
85	1
80	6
75	1

13. In many schools, a grade-point average (GPA) is computed for each student. Usually an A is worth 4 honor points per credit; a B, 3; a C, 2; a D, 1; and an F, 0. To find a grade-point average, do the following:

   (1) Total the number of credits.

   (2) Find the honor points earned for each course by multiplying the honor points for the grade by the number of credits for the course.

   (3) Total the number of honor points.

   (4) Divide the total number of honor points by the total number of credits.

Course	Credits	Grade	Honor Points for Course
Math	1	A	
English	1	B	
Science	1	A	
History	1	C	
Art	0.5	B	
P.E.	0.25	B	
Totals			

   Find the GPA for this student to the nearest hundredth. _____

From *Math for Real Kids* © 2005 David B. Spangler.

Where Do You Draw the Line?

• •

Suppose you are offered two similar jobs. Each job has a starting salary of $26,000. However, at Job A you will receive a $2,000 raise after each year. At Job B, every 6 months you will receive an $800 raise, to be paid over the next 6-month period.

1. Complete the table below to show how much you would earn during each year at each job.

Year	Job A Earnings During Year	Job B Earnings During Year
1	$26,000	
2		
3		
4		

2. Use the information above to make a dual line graph in the grid at the right. Be sure to label which line represents which job.

3. At which job would you earn more money during the first year? _____

4. During which year of work would you earn the same amount at either job? _____

5. Use a ruler to extend each graph. Then answer the remaining questions.

6. Find the difference in earnings between the two jobs during the sixth year of work.

7. How much would you earn with Job B during the eighth year of work?

8. Beginning with year 3 and continuing thereafter, does the difference in earnings between the two jobs increase, decrease, or stay the same?

9. Which job would you take if you planned to stay there less than three years?

Earnings During Each Year at Job A and at Job B

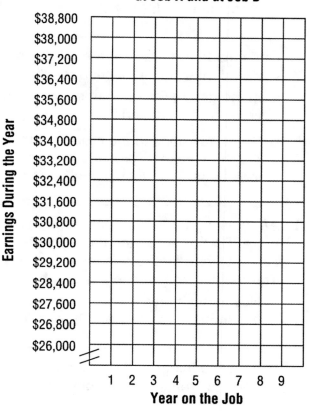

_____

_____

_____

_____

Graphs That Should Be Barred

. . . .

1. This graph shows yearly profits at Termite Lumber Co. from 2002 to 2004.

   a. Have profits been increasing from 2002 to 2004, or has something been "eating away" at the profits?

   _____

   b. What was done to create a *misleading* impression when comparing yearly profits from 2002 to 2004?

   _____

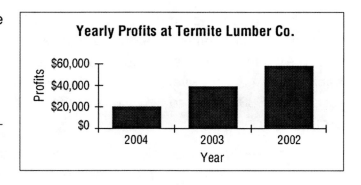

2. Chris made the graph at the right to show the Student Council Election results.

   a. In the graph, Chris's bar is how many times as tall as Gene's? _____

   b. What was done in the creation of the graph to make the results for Chris look better than they actually are?

   _____

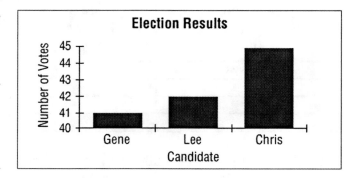

3. After tasting three different brands of pizza (Brands A, B, and C), students were asked to vote for the one they liked best. The results are given in this graph.

   a. How many times as many people voted for Brand A as voted for Brand C?

   _____

   b. Find two things in the graph that were done to create a misleading impression.

   _____

   _____

   c. As a result of factors listed in part **b** above, what illusion has been created by the graph?

   _____

4. Draw the three graphs in this lesson the way they should look. (Use another sheet of paper.)

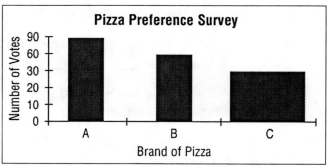

Name: _____ Date: _____

Take Stock of Stocks

....

Stock Market Report for Sonic BoomBoxes, Inc.				
Daily price changes are given in dollars.				
Monday	Tuesday	Wednesday	Thursday	Friday
down $1\frac{1}{4}$	up $\frac{1}{8}$	up $\frac{3}{8}$	up $\frac{7}{8}$	?

Write each answer in simplest form.

1. Find the combined gain for Tuesday and Wednesday.   _____

2. How much greater was Thursday's gain than Tuesday's?   _____

3. Find the combined gain for Tuesday, Wednesday, and Thursday.   _____

4. The price before trading on Monday was $25\frac{3}{4}$ dollars.
   Find Monday's closing price.   _____

5. Thursday's closing price was $25\frac{7}{8}$ dollars. Friday's closing
   price was $23\frac{5}{8}$ dollars. How much did the price per share drop
   on Friday?   _____

6. Tuesday's closing price was $24\frac{5}{8}$ dollars. Find Wednesday's
   closing price.   _____

7. A year ago today, a share sold for $37\frac{3}{8}$ dollars. Today a share
   sells for $25\frac{5}{8}$ dollars. Suppose you bought 1 share at each price.
   How much would you have paid in all?   _____

8. The high price paid this year for a share is $40\frac{1}{2}$ dollars. The low
   price is $19\frac{1}{2}$ dollars. Find the difference between those amounts.

9. Suppose you bought a share for $29\frac{7}{8}$ dollars and sold it for
   $38\frac{1}{8}$ dollars. How much money would you have made?   _____

What Do They Have in Common?

. . . .

The *least common multiple (LCM)* of a set of numbers is the least nonzero number that is a multiple of each number in the set. The least common multiple of 6 and 12 is 12. The least common multiple of 6 and 8 is 24. The least common multiple of 5 and 9 is 45.

1. In a certain video game, on every 6th square you face a *Deeker;* on every 9th square you face an *Eeker;* on every 12th square, a *Queeker;* and on every 15th square, a *Zeeker.* On every how many squares do you face . . .

   a. both a *Deeker* and an *Eeker?*

   b. both an *Eeker* and a *Queeker?*

   c. both a *Queeker* and a *Zeeker?*

   d. a *Deeker,* an *Eeker,* and a *Queeker?*

   _____   _____   _____   _____

2. Rose Greenthumb waters her cactus every 10 days, her fern every 5 days, and her coleus every 3 days. She watered all three plants today. How many days will it be before she waters all three on the same day again?   _____

3. The 17-year cicada comes out every 17 years. The 13-year cicada comes out every 13 years. They both came out at the same time in Illinois in 1868. What is the next year when both will come out again at the same time in Illinois?   _____

4. In a variation of the game Buzz, the leader picks two numbers, say 7 and 8. The first player begins counting with 1. The next player counts 2, and so on. Whenever a multiple of 7 or 8 comes up, a player must say "buzz" instead of the number. However, if the number is a multiple of both 7 and 8, a player must say "buzz-buzz." Find the first "buzz-buzz" number when the following numbers are picked:

   a. 7 and 8 _____   **b.** 18 and 36 _____

   c. 12 and 15 _____   **d.** 21 and 24 _____

From *Math for Real Kids* © 2005 David B. Spangler.

5. On March 3rd, Able and Mable worked in food service. Able works every 4th weekday. Mable works every 5th weekday. Find the next date when both will work on the same day again.

_____

	S	M	T	W	T	F	S
M			1	2	3	4	5
A	6	7	8	9	10	11	12
R	13	14	15	16	17	18	19
C	20	21	22	23	24	25	26
H	27	28	29	30	31		

6. The prices for three brands of pens of the same quality are provided below.

Brand X 4 pens for $5	**Brand Y** 5 pens for $6	**Brand Z** 7 pens for $9

  a. With Brand X, how many pens can you get for $30?   _____

  b. With Brand Y, how many pens can you get for $30?   _____

  c. Which brand, X or Y, offers you more pens for the same money?   _____

  d. Use the idea of *least common multiple* to compare the costs of Brands Y and Z. Explain which of those two brands offers you more pens for the same money.

_____

_____

7. When is the LCM of a set of numbers the largest number in the set?

_____

_____

8. When is the LCM of a set of numbers the product of all the numbers in the set?

_____

_____

Rulers Rule!

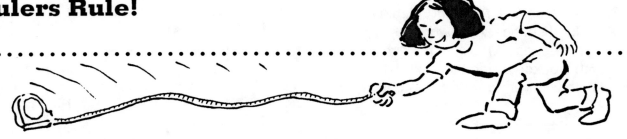

1. The reading for each inch ruler is indicated by an arrow. Write each reading as a fraction or as a mixed numeral in simplest form.

   a.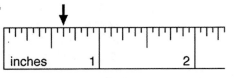

   b.

   a. _____

   b. _____

   c.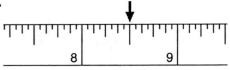

   d.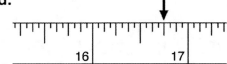

   c. _____

   d. _____

   e.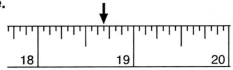

   f.

   e. _____

   f. _____

2. Find, in simplest form, the *sum* of the above readings for...

   a. Rulers **a** and **b** _____   **b.** Rulers **a** and **d** _____   **c.** Rulers **c** and **d** _____

   d. Rulers **c** and **e** _____   **e.** Rulers **d** and **e** _____   **f.** Rulers **b, c,** and **f** _____

3. The *perimeter* of a figure is the distance around the figure.
   Find the perimeter of each figure.

   a.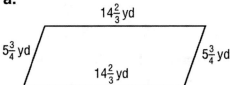

   b.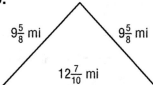

   c.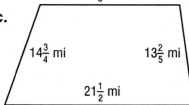

4. The *perimeter* of the triangle at the right is $15\frac{1}{4}$ inches.
   Find the length of the missing side. _____

   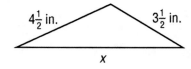

Mixed Numbers at the Olympics

. . . .

Olympic Gold Medal Winners
* indicates Olympic record

Long Jump (Women)			Javelin Throw (Women)		
1984	A. Cusmir-Stanciu, Romania	22 ft 10 in.	1932	Mildred Didrickson, U.S.	$143\frac{1}{3}$ ft
1988	Jackie Joyner-Kersee, U.S.	24 ft $3\frac{1}{2}$ in.*	1936	Tily Fleischer, Germany	$148\frac{11}{48}$ ft
1992	Heike Drechsler, Germany	23 ft $5\frac{1}{4}$ in.	1948	Herma Bauma, Austria	$149\frac{1}{2}$ ft
1996	Chioma Ajunwa, Nigeria	23 ft $4\frac{1}{2}$ in.	1996	Heli Rantanen, Finland	$222\frac{11}{12}$ ft
2000	Heike Drechsler, Germany	22 ft $11\frac{1}{4}$ in.	2000	Trine Hattestad, Norway	?
Long Jump (Men)			Discus Throw (Men)		
1984	Carl Lewis, U.S.	28 ft $\frac{1}{4}$ in.	1984	R. Dannenberg, W. Germany	218 ft 6 in.
1988	Carl Lewis, U.S.	28 ft $7\frac{1}{4}$ in.	1988	Jürgen Schult, E. Germany	225 ft $9\frac{1}{4}$ in.
1992	Carl Lewis, U.S.	28 ft $5\frac{1}{2}$ in.	1992	Romas Ubartas, Lithuania	213 ft $7\frac{3}{4}$ in.
1996	Carl Lewis, U.S.	27 ft $10\frac{3}{4}$ in.	1996	Lars Riedel, Germany	227 ft $8\frac{1}{4}$ in.*
2000	Ivan Pedroso, Cuba	28 ft $\frac{3}{4}$ in.	2000	Virgilijus Alekna, Lithuania	227 ft $4\frac{1}{3}$ in.

Use the above data to solve each problem. Write answers in simplest form.

1. **a.** How much shy of the Olympic record for the men's discus throw was the winning mark in 1984?   _____

   b. Answer the same question for the 1992 winning mark.   _____

   c. In 1996, Lars Riedel beat the men's discus throw record that had been set in 1988. By how much did he beat it?   _____

   d. In 2000, did Virgilijus Alekna miss Lars Riedel's 1996 record by *more than 4 inches* or by *less than 4 inches*? Explain.

   _____

   _____

2. **a.** How much shy of the Olympic record for the women's long jump was the winning mark in 1984?   _____

   b. Answer the same question for the 1996 winning mark.   _____

   c. In which year did Heike Drechsler do better, 1992 or 2000? How much better?   _____ ; _____

3. **a.** The oldest surviving measurement from the ancient Olympic
   Games is a long jump of 23 feet $\frac{1}{2}$ inch by Chionis of Sparta
   in 656 B.C. How much is that shy of Carl Lewis's 1984 jump? _____

   b. How many years ago did Chionis do that jump?
   (Be careful! There is no year 0.)

   c. Carl Lewis's best mark is $7\frac{1}{4}$ inches shy of the Olympic record
   (set in 1968 by Bob Beamon, U.S.).
   What is Beamon's Olympic record? _____

   d. Which two winning marks by Carl Lewis are closest in length
   to each other?

   e. What is the difference in the two marks in part **d** above? _____

   f. By how much did Ivan Pedroso's winning mark beat
   Carl Lewis's *shortest* winning mark? _____

4. **a.** The Olympic record for the women's javelin throw (set in 1988
   by Petra Felke of E. Germany) is $101\frac{2}{3}$ feet longer than the
   1932 winning mark. What is Felke's Olympic record? _____

   b. How much longer is the 1996 winning mark than the 1936
   winning mark? _____

   c. The winning 2000 javelin throw mark for Trine Hattestad is
   missing from the table. However, she beat the 1996 winning
   mark by $3\frac{1}{6}$ feet. What was Trine Hattestad's winning mark
   in 2000? _____

5. The closest race in the history of the Olympic Games was
   decided by $\frac{1}{500}$ th of a second! In a 400-meter individual
   medley, Tim McKee finished in second place with a time of
   4 minutes $31\frac{983}{1000}$ seconds. Give the time for the winner. _____

Fractions and Probability

Spinning into Probability

· ·

Suppose a spinner like this is spun once.

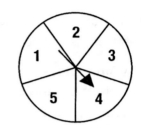

1. **a.** How many possible outcomes are there? _____
    b. Are each of the outcomes as likely to occur as any of
    the others? _____
    Explain why or why not.

    _____

2. **a.** How many of the outcomes are the number 4?   _____
    b. There is _____ chance out of _____ to spin a 4.
    This is your *probability* for spinning a 4.
    c. Write that probability as a fraction.   _____
    d. Write that probability as a decimal. $\frac{1}{5}$ = 1 ÷ 5 =   _____

3. **a.** How many of the outcomes are even numbers?   _____
    b. There are _____ chances out of _____ to spin an even number.
    c. Write that probability as a fraction.   _____
    d. Write that probability as a decimal.   _____

4. **a.** There are _____ chances out of _____ to spin any number
    from 1 through 5.
    b. Write that probability as a fraction.   _____
    c. Write that probability as a whole number.   _____

5. **a.** There are _____ chances out of _____ to spin a number
    greater than 5.
    b. Write that probability as a fraction.   _____
    c. Write that probability as a whole number.   _____

6. **a.** In general, an outcome that will always occur has a probability of _____ .
    b. An outcome that will never occur has a probability of _____ .
    c. Any outcome must have a probability of 0, 1, or a number
    between _____ and _____ .

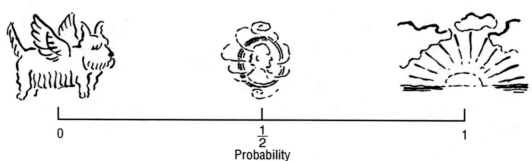

0                    $\frac{1}{2}$ 1
                  Probability

Name: _____   Date: _____

Chance Problems

· ·

The collection of all possible outcomes in an experiment is called a *sample space.*

1. Without looking, you are to draw one card from the hand shown at the right.
   a. How many possible outcomes are in the sample space? _____

   Write, as a fraction in simplest form, the probability of drawing . . .

   b. a spade _____

   c. a club _____

   d. a 5 _____

   e. an 8 _____

   f. a spade or a club _____

   g. a 6 or a king _____

   ♣ clubs   ♠ spades

Suppose you do not know the answer to this multiple-choice question—so you guess.

> **The Boston Tea Party took place in**
> **(a)** 1773   **(b)** 1774   **(c)** 1775   **(d)** 1776

2. What is the probability of guessing . . .
   a. the correct answer? _____   **b.** a wrong answer? _____

3. In answering the above question, suppose you are able to eliminate the choice, 1776. From the remaining three choices, what is the probability of guessing . . .
   a. the correct answer? _____   **b.** a wrong answer? _____

4. The 36 possible sums when two dice are rolled are shown at the right. If a pair of dice is rolled once, what is the probability that the sum will be a . . .
   a. 2 _____   **b.** 3 _____   **c.** 4 _____
   d. 5 _____   **e.** 6 _____   **f.** 7 _____
   g. 8 _____   **h.** 12 _____   **i.** 1 _____

5. You are playing *Monopoly®* and you are on the square "Short Line Railroad." "Park Place" is 2 squares away, and "Boardwalk" is 4 away. On your next roll, what is the probability that you will land on either "Park Place" or "Boardwalk"? _____

Second Die

First Die	1	2	3	4	5	6
1	2	3	4	5	6	7
2	3	4	5	6	7	8
3	4	5	6	7	8	9
4	5	6	7	8	9	10
5	6	7	8	9	10	11
6	7	8	9	10	11	12

| *Fractions and Probability*

Odds Are ...

. . . .

The *probability for* an event is found as follows:

number of outcomes for the event to happen
total number of possible outcomes

The *odds for* an event are expressed as follows:
number of outcomes for the event to happen *to*
number of outcomes for the event not to happen

Converting a probability to odds

A die is rolled once.

- The *probability for* rolling a 3 is $\frac{1}{6}$.
- The *odds for* rolling a 3 are *1 to 5.* This means that in the long run you are likely to roll one 3 for every five times that you don't roll a 3.
- The *odds against* rolling a 3 are *5 to 1.* This means that in the long run, five times in six rolls you are likely *not* to roll a 3. (You are five times as likely not to roll a 3 as you are to roll a 3.)

> Converting a probability to odds:
>
> If the probability for an event is $\frac{a}{b}$,
> then the odds for that event are ***a* to *b − a.***

1. A die is rolled once. Find each probability in simplest form.
   a. probability for rolling an even number _____
   b. probability for not rolling an even number _____
   c. probability for rolling a 1 or 2 _____
   d. probability for not rolling a 1 or 2 _____

2. A die is rolled once. Find each of these odds.
   a. odds for rolling an even number _____
   b. odds for not rolling an even number _____
   c. odds for rolling a 1 or 2 _____
   d. odds for not rolling a 1 or 2 _____

3. One card is drawn from a deck of 52 cards. The probability for drawing a picture card (a jack, a queen, or a king) is $\frac{12}{52}$, or $\frac{3}{13}$. Find the odds . . .
   a. for drawing a picture card _____ **b.** against drawing a picture card _____

Converting odds to a probability

Suppose the odds *for your* winning a game are *2 to 7.* This means that in the long run you are likely to win 2 times for every 7 times you lose. So in 9 games you are likely to win 2 times.

- The *probability for your* winning a game is $\frac{2}{9}$.
- The *odds against* your winning a game are *7 to 2.*

So the *probability against* your winning a game is $\frac{7}{9}$.

> Converting odds to a probability:
>
> If the odds for an event are *a* to *b,*
>
> then the probability for that event is $\frac{a}{a+b}$.

4. Suppose the *odds against* winning a contest are *1,000 to 3.* Find the . . .
   a. probability *against* winning the contest. _____
   b. probability *for* winning the contest. _____

5. Complete this table.

Team	Odds for Winning	Probability for Winning	Odds Against Winning	Probability Against Winning
a. Bears	5 to 2	$\frac{5}{7}$	2 to 5	
b. Packers		$\frac{5}{12}$		
c. Lions	1 to 1			
d. Vikings			2 to 1	

6. A team spokesperson said that the odds are 1 to 2 that the price of tickets will go up next season; 1 to 1 that the price will stay the same; and 2 to 1 that the price will go down. Do the corresponding probabilities "add up"? Explain.

_____

_____

_____

_____

Name: _____ Date: _____

It's Predictable

. . . .

Suppose you roll a die 60 times. You can *predict how* many times you would expect to get a 5 as follows:

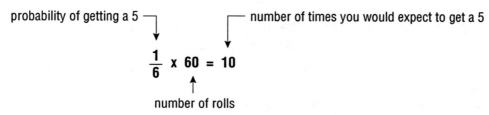

probability of getting a 5 ⌐   ⌐ number of times you would expect to get a 5

$$\frac{1}{6} \times 60 = 10$$

↑ number of rolls

1. A die is rolled 90 times.
   a. Predict how many times you would expect to get a 3.   _____
   b. Predict how many times you would expect to get a 5 or 6.   _____

2. Fifty students were polled to see whom they will vote for in the upcoming student council election. The results are shown at the right.

   a. Find the probability that a student will vote for each candidate.
   b. Then predict how many votes each candidate would expect to get if 750 students vote in the student council election.

Candidate	Votes
Mary Land	21
Della Ware	15
Al Aska	13
Louise E. Ana	1

	Land	Ware	Aska	Ana
Probability				
Prediction				

3. You have run out of time while taking a test—so you blindly answer the question at the right. What is the probability of guessing . . .

   a. the correct answer?   **b.** a wrong answer?

   _____   _____

> **Which amendment gave women the right to vote?**
>
> **(a)** 16th   **(b)** 17th   **(c)** 18th
> **(d)** 19th   **(e)** 20th

4. Suppose there are 20 questions, like the one above, remaining on a test when time runs out. You blindly guess at all 20 questions.

   a. Predict how many points you will get if you get 1 point for each correct answer.   _____

   b. For every wrong answer, $\frac{1}{4}$ point will be deducted. Predict how many points will be deducted if you blindly guess at the 20 questions.   _____

   c. Predict the net result in points if you blindly guess.   _____

5. If a person is selected at random, the probability that he or she was born on leap day (February 29) is $\frac{1}{1,461}$. Predict, to the nearest person, how many people were born on leap day in a city of 100,000.   _____

6. Pharmacists fill about 25,000,000 prescriptions a year. Suppose the probability of getting the correct prescription is $\frac{999}{1,000}$. Predict how many *incorrect* prescriptions would be expected in a year. (Sometimes being correct 999 times out of 1,000 is not good enough.)   _____

IT'S A LONG TIME BETWEEN BIRTHDAYS.

HAPPY BIRTHDAY

Name: _____   Date: _____

What Shape Is He In?

• •

The *tangram* is an ancient Chinese puzzle made by cutting a square into seven special pieces. Each piece is called a tan. For almost 4,000 years, people have enjoyed using tangrams to make various shapes, designs, and objects.

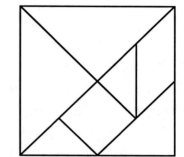

You can make a tangram puzzle by following these steps:

1. Draw a square that is 4 units long on each side. (Use grid paper, or use the grid that is provided below.) Label the large square *ABCD,* as shown below.

2. Complete labeling your square, as shown below.

3. Draw diagonal *BD.* Draw line segments *EF, AJ, HF,* and *JI.*

4. Cut out the seven tangram pieces along the line segments you have drawn. They are triangles *AGD, ABG, DHF, GIJ,* and *FEC;* square *HGJF;* and parallelogram *IBEJ.*

Use your tangram to help you act out the story on the next page.

The following story, Tanaman, was written by Jamie Spangler, a sixth grader from Northbrook, Illinois. She was inspired to write this story after reading *Grandfather Tang's Story* by Ann Tompert. During the story you will be asked to use a set of tangrams (all 7 tans) to make objects and characters that can be used to help act out the events that take place. Enjoy the story!

Tanaman

There once lived a rich man named John who owned an estate in Magicland. But in spite of his riches, John was greedy and foolish. He wasn't satisfied counting his money under a tree. (Make Tangram #1.) He wanted to be the mightiest of all! So John turned himself into a mean crocodile. (Make Tangram #2.) "Crocodiles crawl too much," admitted John, so he decided to become a giraffe. (Make Tangram #3.) However, giraffes can't climb trees, so he morphed himself into a cat. (Make Tangram #4.) But cats can't fly, so he became a bat. (Make Tangram #5.) But he was jealous of swimming animals, so he swooped down into the water and became a fish. (Make Tangram #6.) Suddenly a fierce polar bear appeared! (Make Tangram #7.) "Mercy!" cried John. "You don't show mercy to the poor," accused the bear. "Please," begged John. "All right, I'll let you go," agreed the bear. John made himself a human again, and ran home. (Make Tangram #8. Then make Tangram #9.) From then on he was giving and generous.

Tangram #1	Tangram #2	Tangram #3
Tangram #4	Tangram #5	Tangram #6
Tangram #7	Tangram #8	Tangram #9

Name: _____   Date: _____

Tangrams, Fractions, and Area

In the tangram below, each tan is named with a letter. Also shown is what fractional part each tan is of the entire square.

1. Suppose the area of the tangram is 32 square units. Find the area of each of the seven tans.

Tan	A	B	C	D	E	F	G
Area (in square units)							

2. What is the sum of the areas of tans *A–G* above?   _____

3. Suppose the area of square *D* is 1 square unit. Find the area of the other tans.

Tan	A	B	C	D	E	F	G
Area (in square units)				1			

4. Explain how you could use your tans to verify that each tan is the fractional part of the square as given below.

_____

_____

_____

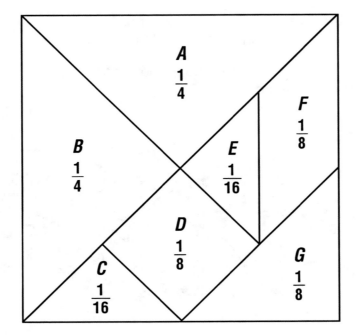

Name: _____   Date: _____

What's Cooking?

· ·

Write each answer in simplest form.

1. To make $1\frac{1}{2}$ times as much as this recipe yields, how much of each of these ingredients will be needed?

   a. _____ cup butter
   b. _____ cup diced onion
   c. _____ ounces tuna
   d. _____ cups diced celery
   e. _____ ounces chow mein noodles
   f. _____ ounces mushroom soup
   g. _____ ounces bean sprouts

2. The new yield will be _____ servings.

3. Use the chart at the right. How long would it take to roast each of these turkeys?

   a. a 6-lb turkey *(Hint:* $6 \times \frac{1}{2} = ?$*)* _____

   b. a $10\frac{1}{2}$-lb turkey _____

   c. a $14\frac{3}{8}$-lb turkey _____

   d. an $18\frac{3}{4}$-lb turkey _____

Turkey Weight (Pounds)	Roasting Time (per Pound at 325°F)
4 to 8	$\frac{1}{2}$ hr
8+ to 12	$\frac{1}{3}$ hr
12+ to 16	$\frac{3}{10}$ hr
16+ to 20	$\frac{1}{4}$ hr

4. When 1 cup of rice is cooked, the yield is 3 cups. Find the yield when $2\frac{2}{3}$ cups of rice are cooked.

   _____

5. A 1-pound box of brown sugar contains $2\frac{1}{3}$ cups. How much sugar is contained in a $1\frac{1}{2}$-pound box?

   _____

6. A 1-pound box of raisins contains $2\frac{1}{2}$ cups. How many cups of raisins are in a $2\frac{3}{4}$-pound box?

   _____

Name: _____   Date: _____

It's Out of This World

· ·

Write each answer in simplest form.

1. **a.** In 1995, the *Endeavor* set a space shuttle record by being in
    space $16\frac{5}{8}$ days. In 1962, John Glenn became the first
    American in orbit. He spent $\frac{5}{24}$ day in space. How much longer
    was the *Endeavor* in space during its record than John Glenn?   _____

    b. How many times as long was the *Endeavor* in space during its
    record than John Glenn?   _____

2. In 1996, U.S. astronaut Shannon Lucid set a record for the greatest number of
    consecutive days in space. Most of her mission was aboard the Russian space station *Mir.*

Shannon Lucid's 1996 Space Mission		
Days in Space	**Miles Traveled**	**Number of orbits**
188	75 million	3,008

    a. About how many months was Shannon Lucid in space?   _____
    b. How many times did she orbit Earth each day?   _____
    c. How long, in hours, did it take to make each orbit?
        Write the answer as a mixed number.   _____

3. In 1999, at age 77, John Glenn became the oldest American to travel in space while aboard the Space Shuttle *Discovery.* He spent about 9 days in space, travelling a total of $3\frac{3}{5}$ million miles. About how many miles did he travel each day? _____

4. The average orbital speed of Venus is 35 kilometers per second. This is $3\frac{1}{2}$ times the average orbital speed of Saturn. Find the average speed of Saturn. _____

5. It takes about $8\frac{1}{3}$ minutes for light to travel 150 million kilometers from the sun to Earth. How many million kilometers does light travel per minute? _____

6. There are $365\frac{1}{4}$ days in 1 *solar year.* A *light-year* is the distance a beam of light travels in 1 solar year. Suppose a beam of light traveled 1,461 days. Give the distance the beam traveled in light-years. _____

7. Our solar system is contained in the Milky Way galaxy. The Milky Way is about 100,000 light-years across. Our solar system is only about $\frac{1}{400}$ light-year across. How many times as long as that is the distance across the Milky Way? _____

8. Mr. Lunar would weigh $25\frac{1}{2}$ pounds on Earth's moon. This is $\frac{1}{6}$ of his weight on Earth. Find his weight on Earth. _____

9. Ms. Pioneer would weigh $321\frac{3}{4}$ pounds on Jupiter. This is $2\frac{3}{5}$ times her weight on Earth. Find her weight on Earth. _____

10. Mars has a diameter that is about half that of Earth, but about twice that of Earth's moon. The diameter of Earth is about 7,900 miles. Use this information to estimate the diameter of Earth's moon. _____

The Spy Who Mixed Numbers

Legend

G – Gate
F – Front Door
R – Refrigerator
B – Back Door

Scale

$6\frac{2}{3}$ yd

Write all answers in simplest form.

A secret spy prowled around the home of Mr. Couchpotato and came out with the scale drawing shown above. Use the drawing to help you answer Questions 1–3.

1. How far is it in yards from the gate to the front door? _____

2. How much farther, in yards, is it from the front door to the refrigerator than it is from the back door to the refrigerator? _____

3. To raid the refrigerator without being seen, the spy figures that she has only $3\frac{1}{3}$ seconds to run from there (R) to the back door. How fast, in yards per second, would she have to run? _____

4. The spy found out that Mr. Couchpotato spends a total of 18 hours each day in his home. Each day he spends $7\frac{3}{4}$ hours eating/watching TV, $8\frac{7}{12}$ hours sleeping, and the rest of the time peeling potatoes. How much time does Mr. Couchpotato spend each day in his home peeling potatoes? _____

The spy used coded letters to send
information back to headquarters.
Part of the code is shown at the right.

			Code		
d	e	f	g	h	i
↓	↓	↓	↓	↓	↓
0	$10\frac{1}{4}$	$10\frac{1}{4}$	$e + f$	$f + g$	$g + h$

5. Use the Code to find each of the following:

   a. $g =$ _____ **b.** $h =$ _____ **c.** $i =$ _____

Use the Decoder at the right to answer the following:

6. How many cookies are there? _____

7. How many cans of cola are there? _____

8. How many bags of chips are there? _____

   (Hint: Extend the pattern in the code to find the number for j.)

Decoder

Number of cookies: $g \div f$

Cans of cola: $i - h$

Bags of chips: $g \times j$

From *Math for Real Kids* © 2005 David B. Spangler.

In My Estimation . . .

....

Some problems do not require an exact answer. A result that is "in the ballpark" may be good enough. We often round numbers or use numbers close to the given numbers in order to make our work easier. When we use such numbers, we are finding an *estimate*.

For Questions 1–6, circle the letter of the best estimate.

1. Your bill at a restaurant is $15.68. You pay with a $20 bill. About how much change should you get back?

   (a) between $3 and $4   **(b)** between $4 and $5   **(c)** between $5 and $6

2. You are being sent to buy 3 hamburgers that cost $2.79 each. About how much money will you need?

   (a) about $6   **(b)** between $7 and $8   **(c)** about $9

3. On a trip you travel 146 miles in 2 hours 57 minutes. Estimate the average speed for the trip.

   (a) about 40 mi/h   **(b)** about 50 mi/h   **(c)** about 60 mi/h   **(d)** about 70 mi/h

4. You quickly punch in the following computation on a calculator:

   $$742 + 1{,}987 + 609 = \boxed{9\ 3\ 3\ 8.}$$

   a. Use estimation to see if your calculator answer is reasonable. The sum should be . . .

   (a) between 2,000 and 3,000   **(b)** between 3,000 and 3,500   **(c)** more than 3,500

   b. Is the calculator answer *reasonable*?   _____

   c. If it is not reasonable, try to figure out what happened.   _____

5. You go shopping and buy items that cost $0.99, $5.48, $12.97, and $8.46. About how much do you spend in all?

   (a) about $25   **(b)** about $26   **(c)** about $28   **(d)** about $29

6. During the past four years, a student has grown the following amounts:

   $1\frac{7}{8}$ in., $2\frac{3}{4}$ in., $3\frac{1}{8}$ in., and $2\frac{15}{16}$ in.

   About how much did the student grow during those four years?

   (a) between 8 in. and 9 in.   **(b)** about $9\frac{1}{2}$ in.   **(c)** between 10 in. and 11 in.

Use "ballpark" estimates to solve these problems. Keep in mind that each has many possible "correct" answers.

7. Estimate the height of a 49-story building if each story is about 13 feet tall.

   _____

8. If a leaky faucet wastes $\frac{1}{20}$ gallon each hour, about how much water will it waste in a week?

   _____

9. In 1790, the U.S. population was 3,929,214. By 1890, it rose to 62,947,714. In 1990, it was 248,709,873.

   a. Estimate the increase from 1790 to 1890.

   _____

   b. About how many times as great was the increase from 1890 to 1990 as it was from 1790 to 1890?

   _____

10. The height of an adult is about 21 times the length of his or her longest finger. Use that "rule of thumb" to estimate the height of an adult whose longest finger is $3\frac{3}{8}$ inches long.

    _____

11. Estimate each probability:

    a. A weather forecaster says, "There is a slightly less than even chance that it will rain today."

    _____

    b. A student says, "It's very likely that I'll do my homework."

    _____

12. Congratulations! You have won a sweepstakes and are given the choice of one of these two prizes:

    • $100 each month for 15 months *OR*

    • Monthly payments of $1, $2, $4, $8, and so on, for 15 months

    a. Which prize has the greater value?

    _____

    b. About how much more money would you get if you chose the prize with the greater value?

    _____

13. Estimate the number of hours you have lived.

    _____

14. Estimate the age of someone who has lived 1 million hours.

    _____

Name: _____   Date: _____

Monster Math

· ·

Scale: 1 cm = 80 km

In Questions 1–3, use a centimeter ruler to find **(a)** each map distance to the nearest tenth centimeter. Then multiply each map distance by 80 to find **(b)** the land distance in kilometers.

1. Mummy's Tomb to
   Wolfman's Den

   a. _____ cm

   b. _____ km

2. Wolfman's Den to
   Dracula's Coffin

   a. _____ cm

   b. _____ km

3. Mummy's Tomb to
   Frankenstein's Castle

   a. _____ cm

   b. _____ km

4. A monster is on the prowl! To find it, begin at Transylvania Station
   and go 240 km south. Then go 360 km northwest. Finally, go
   280 km southwest. The monster is at _____

5. Find **(a)** the map distance from Chicago to
   Denver. Then find **(b)** the actual distance.

   a. _____ (on map)

   b. _____ (actual)

6. What is the actual distance from
   San Francisco to Houston? _____

7. What is **(a)** the map distance from
   Los Angeles to Detroit to New York?
   What is **(b)** the actual distance of
   that trip?

   a. _____ (on map)

   b. _____ (actual)

Scale: 1 cm = 560 km

From *Math for Real Kids* © 2005 David B. Spangler.

Going to Great Lengths

. . . .

The thickness of a dime is about 1 millimeter (mm).

The width of a jumbo paper clip is about 1 centimeter (cm).

10 mm = 1 cm

The length of a golf club is about 1 meter (m).

100 cm = 1 m

If you run around a football field 3 times (including end zones), you will have run about 1 kilometer (km).

1000 m = 1 km

In Questions 1–8, complete each statement with the most reasonable choice from the following units of metric measure:

mm   cm   m   km

1. The diameter of a quarter is 24 _____ .

2. The height of the Sears Tower is 443 _____ .

3. An adult's waist size could be about 85 _____ .

4. The length of a tennis court is 23.77 _____ .

5. The distance from El Paso to Dallas is 995 _____ .

6. In one year, a teenager could grow 60 _____ .

7. The speed limit on many highways is 88 _____ /h.

8. The height of a seventh grader could be about 152 _____ .

9. 1 meter is about 1.09 yards. Which is longer, 1 meter or 1 yard? _____

10. 1 kilometer is about 0.62 mile. Which is longer, 1 kilometer or 1 mile? _____

11. How many kilometers long is the 50,000-meter walk? _____

• •

12. Video cassette tape for a small camcorder is 8 mm wide.
    VHS tape for a VCR (or large camcorder) is 1.25 cm wide.
    How much wider is VHS tape than 8-mm tape? _____

13. The length of an Olympic-sized pool is 50 meters. How many
    lengths would you need to swim to cover 1 kilometer? _____

14. During five days of jogging, Minnie Run jogged a total of 19.5 km.
    Her jogging distances for the first four days are given below:

    3.8 km   4.0 km   4.3 km   4.5 km

    a. Find the average number of kilometers she jogged per day
    during the <u>five</u> days of jogging. _____
    b. How far did she jog on the fifth day? _____

15. In *A Study in Scarlet by* Sir Arthur Conan Doyle, Sherlock Holmes
    tells Dr. Watson the following:

    **Why the height of a man, in nine cases out of ten, can be told
    from the length of his stride. It is a simple calculation enough,
    though there is no use my boring you with figures.**

    Suppose we "bore you with figures." Assume a man's height
    can be found by multiplying his stride by 2.75. Find the height,
    in meters, of a man who has a stride of 60.4 cm. _____

In What Capacity?

. . . .

An eyedropper holds about 1 milliliter (mL) of liquid.

A liter (L) is a little more than a quart.

1000 mL = 1 L

The capacity of $2\frac{1}{2}$ bathtubs is about 1 kiloliter (kL).

1000 L = 1 kL

WHAT DO YOU DO WITH A KILOLITER OF BUBBLES WHEN YOU HAVE ONLY **ONE TUB?**

The amount a container can hold is called its *capacity.*

In Questions 1–5, complete each statement with the most reasonable choice from the following units of metric capacity: **mL   L   kL**

1. A gas tank that is nearly empty needs about 45 _____ .

2. The capacity of a tablespoon is 15 _____ .

3. The daily amount of milk given by a dairy cow is about 30 _____ .

4. A small carton of milk holds 236 _____ .

5. The capacity of a swimming pool could be 60 _____ .

6. An ice tray holds 0.5 liter of water. How many milliliters is that? _____

7. This table shows approximate amounts of water that are used:

Taking a bath: 160 liters	Taking a shower: 15 liters per minute

   a. Which uses more water: a bath or an 11-minute shower? _____

   b. If 365 showers are taken, how many total kiloliters of water would be saved if the showers last 5 minutes each instead of 12 minutes each? _____

8. Your doctor tells you to take 2 teaspoons of cough syrup 3 times a day. How many days will a 150-mL bottle last? (1 teaspoon holds 5 milliliters.) _____

A Weighty Matter

• •

The mass of a flower seed is about 1 milligram (mg).

The mass of a dollar bill is about 1 gram (g).

1000 mg = 1 g

The mass of a liter of water (at 4°C) is 1 kilogram (kg).

1000 g = 1 kg

The water (at 4°C) filling a box 1 meter long on each side has a mass of 1 metric ton (t).

1000 kg = 1 t

Mass is the amount of matter in an object. The terms *mass* and *weight* are often used interchangeably for objects on Earth.

In Questions 1–6, complete each statement with the most reasonable choice from the following:

mg   g   kg   t

1. The mass of a seventh-grade student could be 40 _____ .

2. The mass of a vitamin tablet is about 250 _____ .

3. The mass of a compact car is about 1.35 _____ .

4. The mass of a $\frac{1}{4}$-lb hamburger patty is about 0.1 _____ .

5. The mass of a box of graham crackers is about 250 _____ .

6. The mass of 1 kiloliter of water (at 4°C) is 1 _____ .

7. A *carat* is a unit of mass for precious stones. 1 carat = 200 milligrams.
   a. Find the mass, in milligrams, of a $\frac{3}{4}$-carat diamond. _____
   b. Find the mass, in grams, of a $1\frac{1}{4}$-carat diamond. _____

To a scientist, *weight* means *force.* In the metric system, force is measured in *newtons.* On Earth, the force of gravity on each kilogram of mass is about 9.8 newtons. Use the equation at the right to find the *weight* of an object on Earth.

Force in newtons	Mass in kilograms
↓	↓
$f = 9.8\ m$	

8. **a.** How much will an object with a mass of 90 kilograms weigh (in newtons) on Earth? _____

   b. What is the mass of an object if it weighs 980 newtons on Earth? _____

   c. The force of the moon's gravity is only 0.17 that of Earth's. What is the force of gravity (in newtons) on each kilogram of mass on the moon? _____

9. Use the information below to find the mass for each individual box.

   a. Box A: _____   **b.** Box B: _____   **c.** Box C: _____

156 kilograms

137 kilograms

254 kilograms

Name: _____   Date: _____

Try These—to a Degree

· ·

°C

Water boils 100

90

80

.70

60

50

40

30

20

10

Water freezes 0

-10

°F

220

210 212

200

190

180

170

160

150

140

130

120

110

100

90

80

70

60

50

40

30 32

20

10

1. Give the reading on the Celsius scale at the left.   _____

Complete statements 2–9 by writing the choice (given below the statement) that is most reasonable.

2. The boiling point of water is   _____ °C.
     −20   0   32   100   212

3. The freezing point of water is   _____ °C.
     −20   0   32   100   212

4. Normal body temperature is about   _____ °C.
     37   40   50   98.6   212

5. A person takes a bath in water that is   _____ °C.
     10   25   50   65   80

6. You go sledding when the temperature is   _____ °C.
     −10   10   15   25   32

7. You swim outside when the temperature is   _____ °C.
     10   30   75   90   100

8. You are comfortable camping outdoors when there is an overnight low temperature of   _____ °C.
     15   40   65   80   95

9. The baking temperature of a kitchen oven is   _____ °C.
     20   75   200   1000   2000

This formula can be used to convert a temperature reading from degrees Celsius to degrees Fahrenheit.

This formula can be used to convert a temperature reading from degrees Fahrenheit to degrees Celsius.

degrees Fahrenheit   degrees Celsius
↓   ↓
$f = (1.8 \times c) + 32$

degrees Celsius ⌐— degrees Fahrenheit
↓
$c = \dfrac{f - 32}{1.8}$

10. Use the above formulas to convert these temperatures to the nearest degree.

a. 0°C = _____ °F   **b.** 86°F = _____ °C   **c.** 20°C = _____ °F

d. 99°F = _____ °C   **e.** 35°F = _____ °C   **f.** 2000°C = _____ °F

Here is a quick way you can *estimate* a temperature from one scale, when given a reading from another.

Estimating the Fahrenheit temperature for a given Celsius reading:	**Estimating the Celsius temperature for a given Fahrenheit reading:**
• Multiply the Celsius reading by 2. • Then add 30.	• Divide the Fahrenheit reading by 2. • Then subtract 15.

11. Use the above method to estimate each of these temperature readings.

a. 2°C is about _____ °F.   **b.** 50°C is about _____ °F.

c. 76°F is about _____ °C.   **d.** 38°F is about _____ °C.

This dual line graph shows the daily high and low temperatures for a city during a one-week period.

12. **a.** What was the *range* in temperatures (the difference between the highest and lowest temperatures) for the week?

_____

b. Which three *days* had the largest ranges in temperature?

_____

c. To the nearest degree, find the average high temperature for the week.

_____

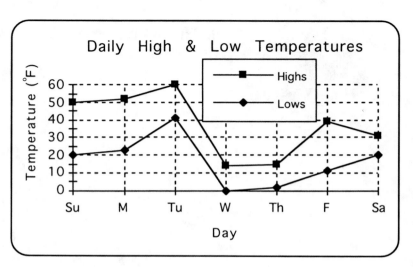

From *Math for Real Kids* © 2005 David B. Spangler.

It's About Time

- -

1. Suppose you have two hours left to spend at a water park. It takes only 10 seconds to go down The Drop water slide. At most, how many times could you go down the slide if each time you have to wait . . .

   a. 35 minutes in line? _____

   b. 5 minutes in line? _____

Train Schedule		
Arrives	**Leaves**	**City**
	7:50 P.M.	Toledo, OH
11:07 P.M.	11:15 P.M.	Cleveland, OH
3:25 A.M.	3:35 A.M.	Buffalo, NY
9:05 A.M.	9:35 A.M.	Albany, NY

Answer Questions 2–4 in hours and minutes.

2. How long must the train travel to get from Toledo to Cleveland? _____

3. How long is the travel time from Cleveland to Buffalo? _____

4. How much longer is the travel time from Cleveland to Buffalo than the travel time from Toledo to Cleveland? _____

5. The train travels 175 miles from Cleveland to Buffalo. Find the average speed of the train during that stretch. (Write the answer from Question 3 as a mixed number before computing.) _____

6. **a.** Complete the time card below.

   b. The employee earns $9.44 per hour. Find the total earnings. _____

Days Worked	Time In	Time Out	Hours Worked
Sunday	1:00 P.M.	4:30 P.M.	3.5
Wednesday	7:00 P.M.	9:15 P.M.	
Saturday	8:30 A.M.	4:15 P.M.	
		Total Hours	

7. Snap your fingers 1 minute from now; wait 2 minutes, do it again; wait twice as long (4 minutes), do it again; and so on. If you continue this, how many times will you snap your fingers in 1 year? _____

This Ratio Is Golden

· ·

A *ratio* is a comparison of two numbers by division. This drawing, including frame, is 7.3 cm by 4.5 cm. The ratio of 7.3 to 4.5 is $\frac{7.3}{4.5}$ or about 1.62.

1. Use a ruler to measure the sides of each rectangle to the nearest tenth of a centimeter. Then find the ratio of the length to the width as a decimal to the nearest tenth.

a.

length: _____
width: _____
ratio of *ℓ* to *w*: _____

b.

length: _____
width: _____
ratio of *ℓ* to *w*: _____

c.

length: _____
width: _____
ratio of *ℓ* to *w*: _____

d.

length: _____
width: _____
ratio of *ℓ* to *w*: _____

e. Which of the four rectangles, if any, look more pleasing to your eyes than the others? _____

Since ancient times, many people have believed that the most pleasing rectangle to look at is one whose ratio of length to width is about 1.62. This is called the *golden ratio*. Rectangles with that ratio are called *golden rectangles*. Such rectangles occur often in art, architecture—and even in the human body!

2. Which of the four rectangles are golden? (If the ratio is close to 1.62, consider it golden.) _____

3. Find the ratio of length to width for each rectangle described below to the nearest hundredth. Then tell if the ratio is golden.

a. a 5" x 8" index card
ratio of *ℓ* to *w*: _____
Is it golden? _____

b. a 3" x 5" index card
ratio of *ℓ* to *w*: _____
Is it golden? _____

c. an $8\frac{1}{2}$" x 11" sheet of paper
ratio of *ℓ* to *w*: _____
Is it golden? _____

Name: _____   Date: _____

A Portion of Proportions

. . . .

A *proportion* is a statement that two ratios (or two rates) are equal.

Example: If a baseball player gets 12 hits in 40 at-bats, you can use a proportion to predict how many hits the player will get in 90 at-bats.

$$\frac{12 \text{ hits}}{40 \text{ at-bats}} = \frac{n \text{ hits}}{90 \text{ at-bats}}$$

The proportion can be solved by finding cross products and then solving for *n*.

$$12 \times 90 = 40 \times n$$
$$1{,}080 = 40 \times n$$
$$n = 27$$

You can predict that the player will get 27 hits in 90 at-bats.

1. You roll a die once. The probability of rolling a 5 is $\frac{1}{6}$. Use a proportion to find how many times you could expect to roll a 5 in . . .

   a. 42 rolls _____   **b.** 72 rolls _____   **c.** 216 rolls _____

For each gear setting on a bicycle, there is a ratio of the number of turns of the pedals to the number of turns of the rear wheel. The ratios appear in this table.

Gear	Number of pedal turns	Number of rear-wheel turns
1st	9	14
2nd	4	7
3rd	1	2
4th	3	7
5th	5	14

2. **a.** While in 2nd gear, Bonnie turned the pedals 20 times. How many times did the rear wheel turn? _____

   b. Joey is in 5th gear. How many times will he have to turn his pedals for his rear wheels to turn 42 times? _____

   c. Ben is in 1st gear. Jamie is in 4th gear. Both turn their pedals 36 times. Whose rear wheel will turn more times? _____

   How many more times? _____ ; _____

UNION TIMES

BLUE BEATS LEVI BY 5 TO 3 MARGIN!

3. **a.** If Blue received 3,000 votes, how many votes did Levi receive? _____

 b. If Levi received 8,715 votes, how many votes did Blue receive? _____

4. The length of a model 747 airplane is 48 centimeters. The scale of the model is 1 to 144. Find the length, in meters, of a real 747 airplane. _____

5. You walk 0.75 mile in 13 minutes on a treadmill. At that rate, how long (to the nearest minute) will it take you to walk 1 mile? _____

For Question 6, round each result to the nearest cent.

ORANGES	APPLES
5 for $1.29	69¢ / lb
GRAPEFRUITS	
Sale Price: 3 for $1.49	
Reg. Price: 2 for $1.49	

6. **a.** How much would it cost to buy 7 oranges? _____

 b. How much would it cost to buy $2\frac{1}{2}$ pounds of apples? _____

 c. How much do you save by buying 6 grapefruits at the sale price instead of at the regular price? _____

The proportion shown below is called the *law of the lever.* Each *w* stands for the weight of an object. The corresponding *d* stands for the distance the object is from the fulcrum. Note that each rate in the proportion compares the weight of one object to the distance of the *other* object.

$$\frac{w_1}{d_2} = \frac{w_2}{d_1}$$

fulcrum

w_1 w_2

d_1 d_2

7. **a.** How far from the fulcrum must a 50-pound object be to balance a 40-pound object located 10 feet from the fulcrum? _____

 b. An 80-pound object is located 9 feet from the fulcrum and balances another object that is located 6 feet from the fulcrum. How heavy is the other object? _____

Percent Potpourri

. . . .

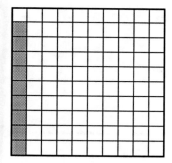

1. **a.** Able said that 9% of the figure is shaded. Gable said that 0.9 of the figure is shaded. Mable said that $\frac{9}{100}$ of the figure is shaded. Fable said that 0.09% of the figure is shaded. Who gave a correct answer?

    b. How many more squares would have to be shaded to cover 75% of the figure?

    _____

A phone company offers the following discounts:

Mon.–Fri.	8:00 A.M. — 5:00 P.M.	Regular Price
	5:00 P.M. — 11:00 P.M.	28% Discount
	11:00 P.M. — 8:00 A.M.	48% Discount
Sat., Sun., and Holidays	All Day	48% Discount

2. The regular price for a certain phone call on Tuesday is $4.00.

    a. How much money would be saved if the call were made at 7:00 P.M. on Tuesday?

    _____

    b. How much would that call cost if it were made on a Sunday?

    _____

3. Last season, football player Art Turf earned $350,000 in pay. Then he received a 120% increase in pay. How much was the increase? _____

4. During the 1995 baseball season, Cal Ripken set a major league record when he played in his 2,131st consecutive baseball game. During that stretch, he played in 99.2% of the innings his team played! If each game was 9 innings long, how many innings did he play during that stretch? Round to the nearest inning. _____

Use these survey results to answer Questions 5–7.

1500 teenage boys and 1500 teenage girls were asked, *"What thing do you most look forward to using your money for when you have completed school and are working?"* The four things mentioned most frequently are given in the table at the right.		Girls	Boys
	Car	20%	47%
	Living away from home	26%	13%
	Clothes	18%	7%
	Contributing to family household	12%	10%

5. What percent of the girls surveyed did not choose one of the top four things listed in the above table?

   _____

   a. How many of the girls surveyed prefer spending money on a car?

   _____

   b. How many of the boys surveyed prefer spending money on a car?

   _____

   c. About what percent of the teenagers surveyed prefer spending money on a car? Choose one of these percents:
   20%   33%   67%

   _____

7. How many of the boys surveyed prefer spending money either on clothes or to help the family household? _____

Name: _____   Date: _____

Building a Solid Foundation on Percents

· ·

The diagram below shows the names and heights of buildings built since 1913 that, when completed, became the *world's tallest building.*

World's Tallest Buildings Since 1913

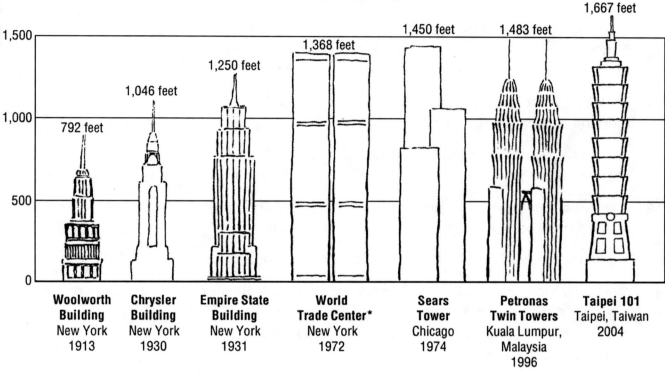

> * The World Trade Center was the world's tallest building from 1972 to 1974. It was destroyed on September 11, 2001.

For Questions 1–3 and 5, find each result to the nearest tenth of a percent.

1. Since 1913, what percent of the seven buildings (or building groupings) that became the world's tallest were in New York? _____

2. What was the *percent increase* in height when . . .
   a. the World Trade Center replaced the Empire State Building as the world's tallest building? _____
   b. the Petronas Twin Towers replaced the Sears Tower as the world's tallest building? _____
   c. Taipei 101 replaced the Petronas Twin Towers as the world's tallest building? _____

3. **a.** After 1913, which building registered the greatest *percent increase* in height when it became the world's tallest building? _____
   b. What was that percent increase in height? _____

4. On July 4, 2004, construction began on Freedom Tower in New York. When completed, it will surpass Taipei 101 in height by about 6.55%. To the nearest foot, how tall will Freedom Tower be? _____

5. **a.** Since 1913, which building has been the world's tallest for the most years? _____

   b. What percent of the time (from 1913 to the present year) has that building been the world's tallest building? _____

6. There are 3 grams of protein in one ounce of *Pro-Teen* cereal. This is 4% of the recommended daily allowance for protein. Find the recommended daily allowance for protein. _____

7. An MP3 player is reduced by $75 to a sale price of $225. Find the percent of discount. _____

8. You buy a computer on sale for $1679.30. This is 30% off the regular price. What is the regular price? _____

9. A merchant paid $15 for a shirt and sold it for $30. The markup was what percent of the price paid by the merchant? _____

For Question 10, round each result to the nearest tenth of a percent.

10. McRoach Motel conducted a survey to see how likely their customers are to come back again. The results are given below.

    a. What percent of those polled are *very likely to* stay at McRoach Motel again? _____

    b. What percent of those polled are either *unlikely or very unlikely* to stay at McRoach Motel again? _____

11. The 1,440 people who completed the McRoach Motel Survey represent only 20% of the people who were given the survey to fill out. How many people were given the survey to fill out?

    _____

McRoach Motel Survey

How likely are you to stay at our motel again?

	number of respondents
(a) very likely	25
(b) likely	39
(c) neither likely nor unlikely	90
(d) unlikely	397
(e) very unlikely	889

Name: _____ Date: _____

Percents That Don't Make Sense

• •

In this lesson you will analyze actual situations where blunders were made involving percents. The names in this lesson have been changed to avoid embarrassing those involved.

This sign, along with several similar to it, appeared in a national office supply store.

1. **a.** How much would this stapler cost if it was priced 100% off list price?

   _____

   b. What is the correct percent reduction in price for the stapler?

   _____

   c. If the stapler really was 714% off list price, how much would the store have to pay you if you "bought" the stapler?

   _____

A table similar to this appeared in a national magazine.

2. **a.** Are the percents in the table reasonable?

   _____

   b. What error was made in each case?

   _____

   c. Give the correct percents (to the nearest percent). 1990: _____ ; 2000: _____

U.S. College Enrollment			
Year	Males	Females	Males as a Percent of the Total
1990	6,284,000	7,535,000	83%
2000	6,722,000	8,591,000	78%

A table of golf fees similar to this appeared in a large city newspaper.

It Costs More to Golf, of Course				
Golf Course	Cost in 1992	Cost in 1996	Amount of Increase	Percent Increase
Village Green	$20	$27	+$7	13.5%
ClubGolf	$45	$90	+$45	20.0%
GolfClub	$9	$30	+$21	33.3%

3. **a.** Explain why the percents in the table are not reasonable.

   _____

   b. Give the correct percent increase for each course (to the nearest percent).
   Village Green _____;   ClubGolf _____;   GolfClub _____

   c. Try to figure out how the percents in the table were computed.

   _____

Name: _____ Date: _____

Take Interest in This

. . . .

Interest is money paid for the use of money.
The amount of money being used is the *principal*.
This formula is used to find *simple interest*:

interest = principal × rate × time (in years)

$$i \quad = \quad p \quad \times \quad r \quad \times \quad t$$

In Questions 1–6, round results to the nearest cent.

Fly-By-Night Savings & Loan	
Type of Account	**Annual Yield***
Passbook	4.61%
1-year	5.81%
4-year	6.90%
6-year	7.17%

*The annual yield is the rate of simple interest when the principal is left on deposit for one full year.

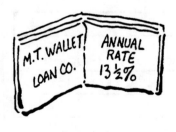

1. Penny opened a passbook account at Fly-By-Night with $800. How much interest will she receive in 1 year?

   _____

2. Dallas Cash opened a 1-year account with $3,500. How much will be in his account after 1 year (interest plus principal)?

   _____

3. Mrs. Moneybags opened a 4-year account with $10,786. The interest will be mailed to her at the end of each year. How much interest will she receive in the 4 years?

   _____

4. If you borrow $5,000 for 1 year from M. T. Wallet Loan Co., how much interest will you have to pay?

   _____

5. Suppose you borrow $7,250 for 2 years. How much would you need to repay the loan (interest plus principal)?

   _____

6. Noah Cashflow borrowed $10,550 for one-half year. How much will he need to repay the loan?

   _____

· ·

The formula below can be used to find the *true annual percentage rate* for any monthly loan plan with 12 or more monthly payments.

$$r = \frac{24c}{p(n+1)}$$

r: **true annual percentage rate**

p: **principal (amount borrowed)**

c: **credit cost (total amount of interest to be paid)**

n: **number of payments**

7. Suppose you buy a $500 TV on credit. You agree to pay $33 a month for 18 months. Find the true annual percentage rate to the nearest percent. Hint: *c* is found as follows:
18 x $33 = $594, so *c* = $594 − $500, or $94.   _____

8. After making a down payment, you owe $8,000 on a car. You will pay $296 a month for 36 months. Find the true annual percentage rate.   _____

Take Even More Interest in This

. . . .

Compound interest is paid on the original principal and on any earned interest that has been left in the account.

In Questions 1, 2, and 4, round each result to the nearest cent.

1. **a.** Suppose you open an account with $1,000 that pays an annual yield of 5.61%. How much interest will you earn in 1 year?   _____

   b. How much will be in your account (interest plus principal) after 1 year?   _____

   c. Suppose you leave the interest and principal in the account for another year. How much interest will you earn during the second year?   _____

   d. Find the total amount in your account after 2 years.   _____

2. An account is opened with $2,500. The annual yield is 7.9%. The principal and all earned interest are left on deposit for 4 full years.

Complete the table at the right to show the principal, interest, and the amount that will be in the account after each of the 4 years. Note: The amount in the account after 1 year becomes the principal for year 2.

Year	Principal	Interest	Amount in Account
1	$2,500.00	$197.50	$2,697.50
2	$2,697.50		
3			
4			

3. The following formula, called the *Rule of 72,* may be used to find the approximate number of years it will take a principal to double when the principal—along with all earned interest—is left in the account.

years to double   **annual yield**
        ↓   ↓
    $y = 72 \div (100 \times r)$

Find the number of years, *y,* it will take a principal to double for each annual yield, *r,* given in the table. Round results to the nearest tenth of a year. (Convert each percent to a decimal when you enter it into the formula.)

	r	*y*
a.	6%	
b.	12%	
c.	10.7%	
d.	3.55%	

4. **a.** Suppose 224 years ago George Washington deposited $1 into an account that paid an annual yield of 6%. If no interest was ever withdrawn, there would now be $466,137.26 in the account. How much interest would the account earn during the 225th year?

   _____

   b. If George Washington *had not* left the interest in the account, he would still have only $1 on deposit. His yearly interest for each of the 225 years would have been only $0.06! How much total interest would he have earned during the 225 years? (Compound interest *pays!*)

   _____

Spread Out Your Spreadsheets

• •

A *spreadsheet* is a table of rows and columns used to organize data. A *computer spreadsheet* is an electronic table used as a powerful tool when sets of calculations must be repeated to solve a problem.

This lesson is written for use with a computer spreadsheet. However, you could also "crunch" the numbers using a calculator and a paper-and-pencil spreadsheet.

Computer spreadsheets usually have numbered rows and lettered columns. The boxes formed by the intersection of the rows and columns are called *cells.* The word "Year" in the spreadsheet below is in cell A1.

	A	B	C
1	Year	Interest	Amount
2	1	$7.00	$107.00
3			
4			

Use a spreadsheet with the directions on these two pages to solve this problem:
$100 is deposited into an account with an annual yield of 7%. The $100 and all interest are left on deposit. How much money will be in the account after 10 years?

1. In your spreadsheet, type "Year" in cell A1, "Interest" in cell B1, and "Amount" in cell C1, as shown above. After each entry, press the RETURN key.

2. Type "1" in cell A2, as shown above.

3. Because we will be finding compound interest, the entries in the "Interest" and "Amount" columns should be in dollars and cents. To do this, click on "B" at the top of the spreadsheet and drag to "C." Columns B and C should be entirely highlighted. For most spreadsheet programs, go to the *Format* menu and choose *Number.* Choose *Currency* to show numbers to two decimal places.

4. To find the interest for the first year, type this formula in cell B2: **=0.07*100.** An "=" sign tells the computer that a formula is coming. The "*" means *multiply.* So this formula tells the computer to find the product 0.07 x 100. The computer will display the product, $7.00, in cell B2. (It will not display the formula.)

5. To find the total amount in the account after 1 year, type this formula in cell C2: **=B2+100.** This will add the result in B2 (which is $7.00) to $100. The sum, $107.00, is displayed in cell C2.

6. Because you will be doing calculations for 10 years, list the years from 1 to 10 in column A. Here is a fast way to do this:

   Type this formula in cell A3: **=A2+1**. This will add the result in A2 (which is 1) to 1. The computer will display the sum, 2, in cell A3.

   Then click on cell A3 and drag down to cell A11. Go to the *Calculate* menu (or *Edit* menu for some spreadsheet programs) and choose *Fill Down.* This will copy and update the formula from one cell to the next.

   Shown below at the left is what the computer will display for the first 3 years. Shown below at the right are the formulas that are in each cell.

What the computer displays

	A	B	C
1	Year	Interest	Amount
2	1	$7.00	$107.00
3	2		
4	3		

Formulas in each cell

	A	B	C
1	Year	Interest	Amount
2	1	=0.07*100	=B2+100
3	=A2+1		
4	=A3+1		

7. You will now write a formula to show the interest for Year 2. Because we need to find 7% of the amount in cell C2, type this formula in cell B3: **=0.07*C2**

8. To find the total amount in the account after 2 years, type the following formula in cell C3: **=B3+C2**

9. Do all of the computations for the remaining 8 years by using the *Fill Down* feature as follows: Click on cell B3 and drag to cell C3. Without lifting your finger from the mouse, drag down to cell C11. Go to the *Calculate* menu (or *Edit* menu) and choose *Fill Down.*

The computer display for Years 1–3 is below left; the formulas, below right.

What the computer displays

	A	B	C
1	Year	Interest	Amount
2	1	$7.00	$107.00
3	2	$7.49	$114.49
4	3	$8.01	$122.50

Formulas in each cell

	A	B	C
1	Year	Interest	Amount
2	1	=0.07*100	=B2+100
3	=A2+1	=0.07*C2	=B3+C2
4	=A3+1	=0.07*C3	=B4+C3

10. How much will be in the account after 10 years? _____

11. Extend your spreadsheet to find out how long it will take the $100 principal to triple. How long will it take? _____

Choose a Calculation Method

• •

When you solve a problem, you probably use one or more of these methods:

• **Estimation.** You use estimation when it is not necessary to find an exact answer.

• **Mental Math.** You use mental math when you need to compute an exact answer in your head.

• **Paper and Pencil.** You use paper and pencil when the computations are too difficult to do in your head or when you need a record of your work.

• **Calculator.** You use calculators when they are available and when the computations are too time-consuming or involved to do by hand or to do mentally.

For each problem, **(a)** indicate which method (or methods) you would use to solve the problem, **(b)** explain why you would use that method, and **(c)** solve the problem using your method. (Most problems could be solved by more than one possible method.)

1. As a cola vendor at a ball game, you have a large basket of cups filled with cola strapped to you. A customer gives you a $20 bill for a $6.75 purchase. How much change should you give back?

a. Calculation Method:	**b.** Why did you choose this method?
c. Answer:	

2. You are a clerk in a clothing store that is having a 35% off sale. Your manager asks you to change all the price tags to show the new prices. You begin with the $169.99 suits. What will be the new price for these suits?

a. Calculation Method:	**b.** Why did you choose this method?
c. Answer:	

From *Math for Real Kids* © 2005 David B. Spangler.

3. You are in a souvenir shop and would like to buy a T-shirt for $12.99, a calendar for $5.75, and 6 postcards at 35¢ each. You have $20. Do you have enough money?

a. Calculation Method:	**b.** Why did you choose this method?
c. Answer:	

4. You are traveling in a car at 55 miles per hour. A sign indicates that Joey Town is 88 miles away. About how long will it take you to get to Joey Town?

a. Calculation Method:	**b.** Why did you choose this method?
c. Answer:	

5. You are baking at home. A recipe calls for $5\frac{1}{4}$ cups of flour and $1\frac{1}{2}$ cups of nuts. How much of each ingredient should you use to make only one-fourth of the recipe?

a. Calculation Method:	**b.** Why did you choose this method?
c. Answer:	

6. You are eating in a restaurant and would like to leave a 15% tip on a $12.76 bill. How much tip should you leave?

a. Calculation Method:	**b.** Why did you choose this method?
c. Answer:	

7. You and three friends plan to have a round-robin table tennis tournament. Each of you will play each other exactly once. How many games will be played in all?

a. Calculation Method:	**b.** Why did you choose this method?
c. Answer:	

What's Inside a Rectangle?

. . . .

The *area* of a figure is the number of square units inside the figure. The area of a rectangle is found by multiplying the measure of its length and the measure of its width. The symbol for square feet is ft², for square meters, m², and so on.

1. Each tile in the above drawing is a square that is 1 foot long on a side. What is the area of the large rectangle? _____

2. Rectangle *A* is 6 feet by 4 feet. Rectangle *B* is 5 feet long on each side. How much greater is the area of rectangle *B* than rectangle *A*? _____

3. Give the dimensions of two rectangles that have the same area, but different perimeters (distances around). _____; _____

4. Give the dimensions of two rectangles that have the same perimeter, but different areas. _____; _____

5. A soccer field has an area of 7,350 square meters. The length is 105 meters. Find the width. _____

6. An *acre* is 4,840 square yards—almost the size of a football field, excluding end zones. How many acres of land does a farmer have if his rectangular field is 363 yards by 100 yards? _____

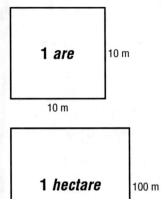

1 are   10 m

10 m

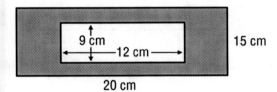

1 hectare   100 m

100 m

7. a. An *are* is a unit of area in the metric system. The square at the left has an area of 1 *are.* How many square meters are in 1 *are?*

    _____

   b. How many square meters are in 100 *ares?*

    _____

8. a. The square at the left has an area of 1 hectare. How many square meters are in 1 hectare?

    _____

   b. How many *ares* are in 1 hectare?

    _____

   c. There are 100 hectares in 1 square kilometer. How many *ares* are in 1 square kilometer?

    _____

9. Find the area of the shaded region in the figure below.   _____

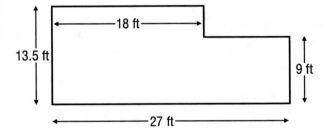

9 cm

—12 cm—

15 cm

20 cm

10. How many square yards of carpet would you need for the room shown below? (9 ft² = 1 yd²)   _____

—18 ft—

13.5 ft

9 ft

—27 ft—

Name: _____   Date: _____

Base Times Height—and You'll Be Right!

. . . .

Area Formulas		
Parallelogram	**Triangle**	**Trapezoid**
$A = b \times h$	$A = \frac{1}{2} \times b \times h$	$A = \frac{1}{2} \times (b_1 + b_2) \times h$
Area = base x height	Area = $\frac{1}{2}$ x base x height	Area = $\frac{1}{2}$ x (sum of the bases) x height

In shuffleboard, players use a cue to slide discs from one end of the court to the other. The goal is to score as many points in the numbered regions as you can. Part of a shuffleboard court appears at the right.

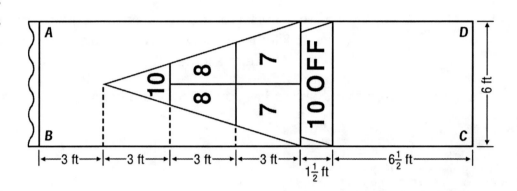

1. Find the area of rectangle *ABCD*. _____

2. The base of the 10-point triangle is 2 feet. The height is given in the above diagram. Find the area of the 10-point triangle. _____

3. The top base of each 8-point trapezoid is 1 foot. The bottom base for each is 2 feet. Find the area of each 8-point trapezoid. _____

4. The top base of each 7-point trapezoid is 2 feet. The bottom base for each is 3 feet. Find the area of each 7-point trapezoid. _____

5. The top base of the 10-OFF trapezoid is 5 feet. The bottom base is 6 feet. (The little triangles at either side of the 10-OFF region are not part of the 10-OFF region.) Find the area of the 10-OFF trapezoid. _____

• •

Let's assume that when you slide a disc (from the left side of the court), it will land in rectangle *ABCD*. You could find the *probability* that the disc will land in one of the numbered regions by dividing the area of the numbered region by the area of rectangle *ABCD*. This type of probability is called *geometric probability*.

6. A disc is slid and lands in rectangle *ABCD*. Find the geometric probability to the nearest thousandth that it lands in . . .

   a. the 10-point triangle. _____

   b. an 8-point trapezoid. (Note: Because there are two 8-point trapezoids, multiply the geometric probability for one trapezoid by 2.) _____

   c. the region of rectangle *ABCD* that is not numbered. _____

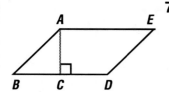

7. **a.** *ABDE* is a parallelogram. If triangle *ABC* is slid so that line segment *AB* is on top of line segment *ED,* the resulting figure will be a _____

   b. Find the area of parallelogram *ABDE*. _____

8. The area of a parallelogram is $13\frac{3}{4}$ square feet. Its base is $5\frac{1}{2}$ square feet. Find its height. _____

9. The area of a triangle is 100 square yards. Its height is 10 yards. Find its base. _____

10. The area of rectangle *AEFD* below (Figure 1) is 20.54 m². Find the area of parallelogram *ABCD*. Note: Line segment *AE* is a height of both the rectangle and the parallelogram. _____

11. The area of parallelogram *JKLM* below (Figure 2) is $20\frac{3}{4}$ in². Find the area of trapezoid *JKPQ*. _____

12. Find the area of the shaded square in Figure 3. _____

Figure 1

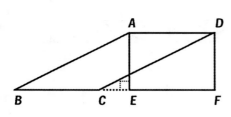

Figure 2

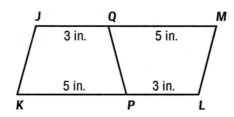

Figure 3

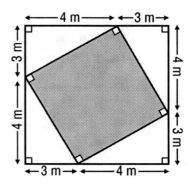

Name: _____   Date: _____

Going Around in Circles

. . . .

Circle Formulas (Use 3.14 for π.)	
Circumference	**Area**
$C = 2 \times \pi \times r$ or $C = \pi \times d$	$A = \pi \times r^2$
Circumference = 2 × π × radius, or π × diameter	Area = π × radius × radius

1. In one revolution of a wheel, a bicycle travels the circumference
   of the wheel. How far does a 24-inch bicycle (24-inch wheel
   diameter) travel when the wheel makes one complete revolution? _____

2. **a.** How far will a car tire with a 13-inch *radius* travel in
   one revolution?   _____

   b. Suppose the tire is worn down 0.25 inch to a radius of 12.75
   inches. Now how far will the tire travel in one revolution?   _____

   c. Do you get better or worse gasoline mileage when your
   tire is worn?   _____

   d. Explain your answer to **c.** _____

3. A radar screen shows a circular region with a radius of 100 miles.
   Find the *area* of the region.   _____

4. The circle appears in the large 10-by-10 square. Each ⬚ is one square unit.

   a. Estimate the area of the circle by
   counting squares and partial squares
   inside the circle.

   _____

   b. Find the area of the circle by using the
   formula.

   _____

   c. Use your result from **4b** to find the
   area of the large square that is NOT
   inside the circle.

   _____

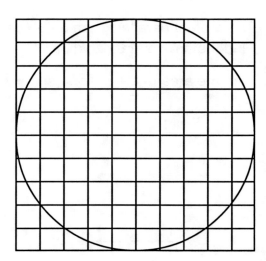

5. At Leaning Tower of Pizza, one 14-inch pizza (*diameter* of 14 inches) costs the same as two 7-inch pizzas.

   a. Which gives you more pizza, one 14-inch pizza or two 7-inch pizzas?

   b. How much more?

6. In Figure 1 below, how much greater is the circumference of the outer circle than that of the inner circle?

7. In Figure 2 below, estimate the area of the outlined region in square units.

8. In Figure 2, estimate the *arc length* (distance around the curved path).

9. In Figure 3, find the distance around the figure.

10. In Figure 3, find the area of the region.

Figure 1

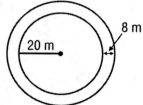

20 m   8 m

Figure 2

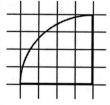

Figure 3

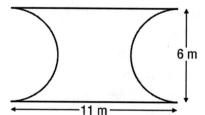

6 m

11 m

Name: _____   Date: _____

Turn Up the Volume

• •

Volume Formulas			
Rectangular Prism (or cube) $V = \ell \times w \times h$ Volume = length x width x height	**Square Pyramid** $V = \dfrac{s^2 \times h}{3}$ Volume = (side x side x height) ÷ 3	**Cylinder** $V = \pi \times r^2 \times h$ Volume = π x radius x radius x height	**Cone** $V = \dfrac{\pi \times r^2 \times h}{3}$ Volume = π x (radius x radius x height) ÷ 3

Volume is given in *cubic units.* The symbol for cubic feet is ft³, the symbol for cubic meters is m³, and so on.

You can use grid paper to make a box. Cut squares away from the corners of the paper, and fold the sides to form a box (without a lid).

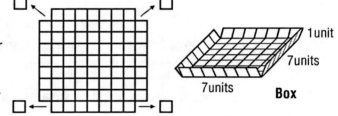

1. **a.** One square has been cut from each corner of the piece of 9-by-9 grid paper shown at the right. The volume of the resulting box is 7 x 7 x 1, or _____ cubic units.

   b. Suppose a 2-by-2 square is cut from each corner of a 9-by-9 grid. Find the volume of the resulting box.   _____

   c. Suppose a 3-by-3 square is cut from each corner of a 9-by-9 grid. Find the volume of the resulting box.   _____

2. Gina dug a hole in the shape of a cube. Each side was 5 feet long. Tina dug two cubic holes, each 2.5 feet on a side. Did they dig the same amount of dirt? If not, explain.

   _____

3. **a.** Find the volume of each cereal box shown at the right.

   15-oz box _____   18-oz box _____

   b. Refer to the weight of the cereal in each box. Which box contains more cereal per cubic centimeter of space—and hence, less air? _____

From *Math for Real Kids* © 2005 David B. Spangler.

4. The Pyramid of Khufu, called the "Great Pyramid," has a square base of 720 feet on a side. It had an original height of 481 feet. Find the original volume of the pyramid. _____

5. A square pyramid is 6.6 meters long on a side and has a height of 6.6 meters. A cube is 6.6 meters long on a side. How many times as great is the volume of the cube than that of the pyramid? _____

6. Find the volume of the cola can at the right. _____

7. Find the volume of the corn can at the right. _____

8. Find the volume of the tuna can at the right. _____

9. Compare the cola can and the corn can. Does doubling the *height* of a can give you twice the volume? _____

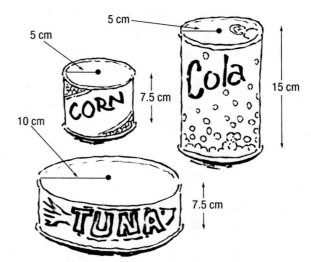

5 cm

5 cm

7.5 cm

15 cm

10 cm

7.5 cm

10. Compare the tuna can and the corn can. Does doubling the radius of a can give you twice the volume? Explain. _____

11. A cone and a cylinder each have a radius of 2 inches and a height of 3 inches. How many times as great is the volume of the cylinder? _____

U-Scream Ice Cream is packaged as shown.

12. Find the volume of the ice-cream bar at the right.

_____

13. Find the volume of the portion of the cylinder that contains ice cream.

_____

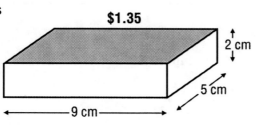

$1.35

2 cm

5 cm

9 cm

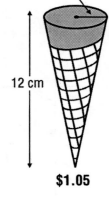

2 cm

7 cm

2.6 cm

12 cm

$1.14

$1.05

14. Find the volume of the cone to the nearest tenth cubic centimeter. _____

15. Which of the three gives the most ice cream for your money? _____

Estimation with Area and Volume

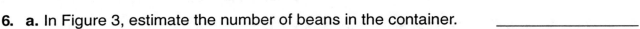

1. A farmer hopes to harvest 13,100 bushels of shell corn from 109 acres of land. About how many bushels is she hoping to harvest per acre?   _____

2. One liter of paint will cover about 13 square meters. About how many liters of paint are needed for a rectangular ceiling 4.8 meters by 7.9 meters?   _____

3. To estimate the weight of objects, movers figure an average of 7 pounds per cubic foot. Estimate the combined weight of the following objects: _____

      sofa, 25.1 cubic feet   refrigerator, 39 cubic feet
      bed, 24.9 cubic feet   4 chairs, 2.5 cubic feet each

4. In Figure 1 below, each ☐ is one square unit. Estimate the area of the outlined region in square units.   _____

5. In Figure 2, estimate the ratio of the following measures of area: Area of Triangle ABC to the area of the circle to the area of the square.
_____ to _____ to _____

6. a. In Figure 3, estimate the number of beans in the container.   _____
   b. Describe how you arrived at your estimate.

_____

Figure 1   **Figure 2**   **Figure 3**

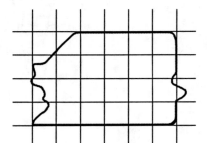
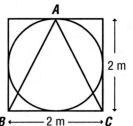
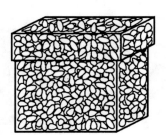

The area of an irregularly shaped region, like the one at the right, can be estimated by separating it into rough trapezoids. The region at the left has been separated into four rough trapezoids, each with a height of 9 millimeters. The bases are given.

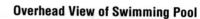
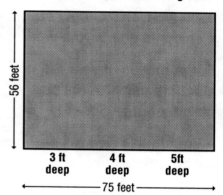

25 mm
36 mm
49 mm
19 mm
13 mm

7. Estimate the area of the region by finding the sum of the area measures of the four trapezoids.

_____

8. **a.** Estimate the volume of the swimming pool at the right in cubic feet.

   _____

   b. Use another piece of paper to explain how you arrived at your estimate.

   c. There are 7.481 gallons in 1 cubic foot of water. Estimate the capacity of the pool in gallons.

   _____

   d. One gallon of water weighs 8.3 pounds. Estimate the weight of the water in the pool.

   _____

Overhead View of Swimming Pool

56 feet

3 ft
deep

4 ft
deep

5ft
deep

75 feet

Hidden Figures

• •

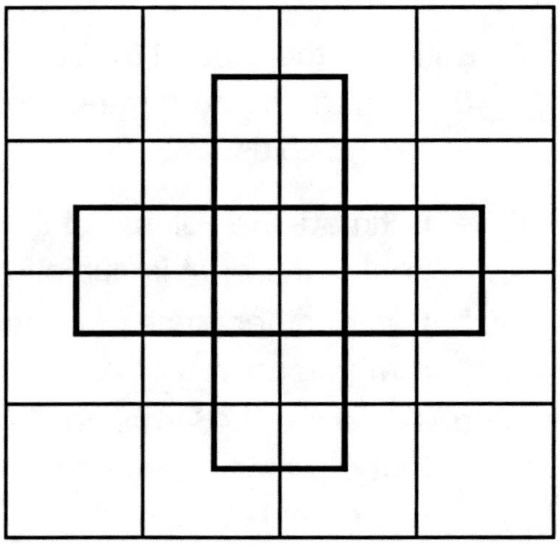

How many squares can you find in the figure at the right?

Before you plunge into the problem, you may want to consider using a *systematic approach*. You could make a list showing how many squares there are of each type. This should save a lot of time—and make it easier to check your work.

For purposes of your list, a square this size will be called a *Large Square*.

A square this size will be called a *Small Square*.

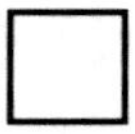

1. Count how many squares there are of each type and record the results in the table below. Note: A 2-by-2 square is 2 squares long on each side. A 3-by-3 square is 3 squares long on each side, and so on.

List of Squares

Number of squares that are a 1-by-1 Large Square: _____

Number of squares that are a 2-by-2 Large Square: _____

Number of squares that are a 3-by-3 Large Square: _____

Number of squares that are a 4-by-4 Large Square: _____

Number of squares that are a 1-by-1 Small Square: _____

Number of squares that are a 2-by-2 Small Square, where the square is completely bounded by a heavy border: _____

Number of squares that are a 2-by-2 Small Square, where the square has exactly two sides bounded by a heavy border: _____

Other (mention type): _____

Total Number of Squares: _____

---

2. How many triangles can you find in the
figure at the right? _____

Once again, it is wise to make a list. One method
is to begin counting (and listing) all triangles that
have a vertex at *A.* Then list all triangles that have
a vertex at *B*—that don't also have a vertex at *A.*
Then continue with all triangles that have a vertex
at *C* and haven't already been listed. Continue
until you cannot find any more triangles.

Good luck.

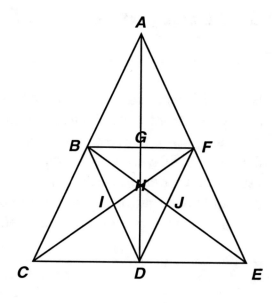

List of Triangles

_____

_____

_____

_____

_____

_____

_____

_____

_____

Tackling Integers I

· ·

Here is how you can use a football field to help you add integers. An addition problem consists of two plays—one for each addend. Begin on the 0-yard line. A positive addend represents a *gain* in yardage (movement to the right). A negative addend represents a *loss* in yardage (movement to the left).

Integer Problem	Pair of Related Football Plays	Illustration	Net Result
-2 + 4	a loss of 2 yards; a gain of 4 yards.	-2 -1 0 1 2	2
4 + -5	a gain of 4 yards; a loss of 5 yards.	-1 0 1 2 3 4	-1

Use this football field to help you find each net result.

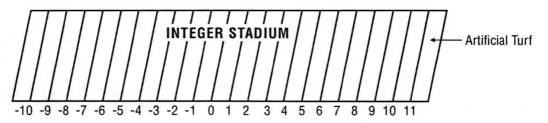

INTEGER STADIUM — Artificial Turf

-10 -9 -8 -7 -6 -5 -4 -3 -2 -1 0 1 2 3 4 5 6 7 8 9 10 11

1. 6 + -4 = _____   **2.** -5 + 3 = _____   **3.** -8 + -2 = _____

4. 9 + -10 = _____   **5.** -7 + 11 = _____   **6.** -6 + 6 = _____

7. 11 + -7 = _____   **8.** 0 + -9 = _____   **9.** -5 + -5 = _____

10. A team lost 3 yards on first down (play) and gained 7 yards on second down. Find the net result.   _____

11. The SubZeroes lost 1 yard on first down, 2 yards on second down, 3 yards on third down, and 4 yards on fourth down. Find the net result for the 4 downs. (To prepare for its games, this team spends 5 hours each week studying negative numbers.)   _____

Tackling Integers II

• •

A subtraction problem can be viewed as two football plays—one for the minuend (first number) and one for the subtrahend (number being subtracted).

• Begin on the 0-yard line. Move the ball to the right if the minuend is positive. Move it to the left if it is negative.

• The second play is a penalty. If the subtrahend is positive, the penalty is *against you* (movement to the left). If the subtrahend is negative, the penalty is *against the opposing team* (movement to the right).

Integer Problem	Pair of Related Football Plays	Illustration	Net Result
2 – 5	a gain of 2 yards; a 5-yard penalty against you	-3 -2 -1 0 1 2	-3
-2 – -5	a loss of 2 yards; a 5-yard penalty against the other team.	-2 -1 0 1 2 3	3

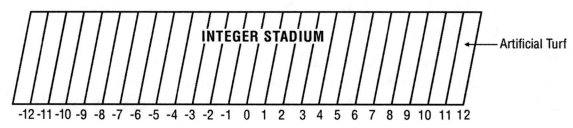

INTEGER STADIUM   ← Artificial Turf

-12 -11 -10 -9 -8 -7 -6 -5 -4 -3 -2 -1 0 1 2 3 4 5 6 7 8 9 10 11 12

1. -4 – 2 = _____   **2.** -3 – -5= _____   **3.** 8 – 12 = _____

4. -7 – -7= _____   **5.** -3 – 5= _____   **6.** 1 – -10= _____

7. -11 – -6 = _____   **8.** 0 – -9 = _____   **9.** 9 – 20 = _____

10. A team lost 6 yards on first down. On second down the opposing team was penalized 10 yards. Find the net result for the two downs. _____

11. The Illegal Procedures gained 11 yards on first down. However, they were penalized 5 yards on second down, 10 yards on third down, and 15 yards on fourth down. Find the net result for the four downs. _____

Name: _____   Date: _____

Wind Chill Numb-brrs

· ·

Wind chill is an estimate as to how cold the body *feels* when considering the combination of both cold *and* wind. For example, suppose the actual temperature (without wind) is 10°F. When this is combined with a 10-mile-per-hour wind, it *feels like* it is –9°F (as shown in the table below).

Wind Chill Table				
Wind Speed	Actual Temperature			
0 mi/h	10°F	0°F	-10°F	-20°F
10 mi/h	-9	-22	-34	-46
15 mi/h	-18	-31	-45	-58
20 mi/h	-24	-39	-53	-67

1. With no wind, the temperature outside is 0°F. Suddenly Mr. O. M. Winter blows a gust of wind that makes it feel 39 degrees colder than that. Use the table to find the speed of the wind. _____

2. Suppose with no wind the temperature outside is 10°F. How many degrees colder would it feel with a wind of . . .

    a. 20 mi/h? _____ **b.** 15 mi/h? _____ **c.** 10 mi/h? _____
    Hint: 10 – -24 = ?

3. Suppose with no wind the temperature outside is -10°F. How many degrees colder would it feel if there suddenly was a wind speed of . . .

    a. 20 mi/h? _____ **b.** 15 mi/h? _____ **c.** 10 mi/h? _____

4. Suppose you are told that the wind speed is 20 miles per hour and that the wind chill temperature is –60°F. Use the table to estimate what the actual temperature would be if there was no wind. _____

5. With no wind, the temperature outside is 5°F. Suddenly, Mr. Winter blows a gust of wind that makes it feel 45 degrees colder than that. Find the resulting wind chill temperature. _____

6. One morning the wind chill temperature was –7°F. When Mr. Winter got tired of blowing wind, the temperature felt 23 degrees warmer. Find the resulting temperature. _____

Name: _____   Date: _____

Positively Negative

• •

Your school sends a team of students to compete in an academic tournament against a team from School *X* and a team from School *Y*. A team scores 15 points for each question it answers correctly and -9 points for each question it answers incorrectly. When a team "passes" (chooses not to answer), 0 points are scored. The results are given in the table below.

School	Correct Answers	Wrong Answers	Passes	Total Score
Your School	16	7	7	
School X	10	18	2	
School Y	5	19	6	

1. Complete the table to find the total score for each school.

2. By how many points did your school defeat its closest rival? _____

3. By how many points did School *X* outscore School *Y*? _____

4. Last year a person went on a diet, resulting in the monthly changes in weight (given in pounds) shown below.

   -8   -6   -3   -1   0   +4   +6   +10   +1   -7   -9   -11

   a. Find the person's net change in weight for the year. _____
   b. Find the *average* change in weight per month. _____

5. The Red Ink Company bought some merchandise for $75, sold it for $65, bought it back for $55, and sold it for $45. Find the net result for that series of transactions. _____

6. Use the formula at the right to convert these very cold Fahrenheit temperatures to Celsius temperatures (to the nearest tenth of a degree).

degrees Celsius	degrees Fahrenheit
↓	↓
$c = \dfrac{f - 32}{1.8}$	

   a. lowest temperature ever recorded on Earth (Antarctica): $-128.6°F =$ _____ °C
   b. coldest place in the solar system (Triton, a Neptune moon): $-391°F =$ _____ °C
   c. absolute zero (lowest possible temperature in the universe): $-459.6°F =$ _____ °C

Getting to the Root of the Problem

The *square root* of a number, *n,* is a number which, when multiplied by itself, gives *n.* The square root of 16 (written $\sqrt{16}$) is equal to 4, because 4 x 4 = 16. However, before calculators with square root keys were available, it was not easy to find the square root of, say, 3. The flowchart below provides a way to estimate the square root of a number to the nearest hundredth. A *flowchart* shows a sequence of steps that need to be followed to perform a task.

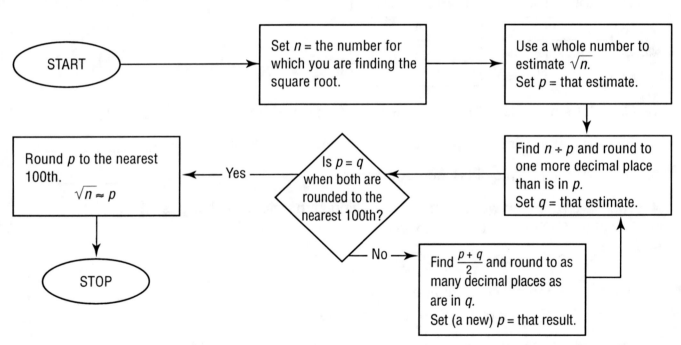

START → Set *n* = the number for which you are finding the square root. → Use a whole number to estimate $\sqrt{n}$. Set *p* = that estimate. → Find *n* ÷ *p* and round to one more decimal place than is in *p.* Set *q* = that estimate. → Is *p* = *q* when both are rounded to the nearest 100th? → Yes → Round *p* to the nearest 100th. $\sqrt{n} \approx p$ → STOP → No → Find $\frac{p+q}{2}$ and round to as many decimal places as are in *q.* Set (a new) *p* = that result.

1. Use the flowchart to estimate each square root to the nearest hundredth.

   a. $\sqrt{3} \approx$ _____   **b.** $\sqrt{7} \approx$ _____   **c.** $\sqrt{11} \approx$ _____

   d. $\sqrt{30} \approx$ _____   **e.** $\sqrt{65} \approx$ _____   **f.** $\sqrt{143} \approx$ _____

This formula shows how to find how far you can see, from horizon to horizon, from the window of a jet plane.

2. Find the view to the nearest hundredth mile for each of these jet altitudes.

> **view in miles**   **altitude of jet in feet**
>
> $v = 1.22 \times \sqrt{a}$

   a. 1,000 ft _____   **b.** 5,000 ft_____   **c.** 30,000 ft_____

Name: _____   Date: _____

Pythagorean Problems

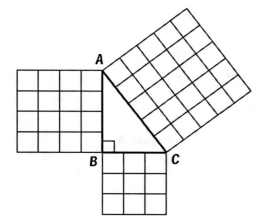

1. Find the area in square units for each of the following:
   a. square with side $\overline{AB}$ _____
   b. square with side $\overline{BC}$ _____
   c. square with side $\overline{CA}$ _____

2. Describe a relationship among the areas of the squares. _____
   _____
   _____
   _____

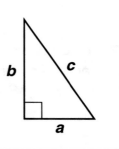

Pythagorean Rule

Suppose a right triangle has legs with lengths *a* and *b*. Suppose the *hypotenuse* (side opposite the right angle) has a length *c*. Then

$$a^2 + b^2 = c^2.$$

About 2,500 years ago, the Greek mathematician Pythagoras proved that this rule is true for *all* right triangles.

3. The size of a TV screen is the length of its diagonal (distance from an upper corner on one side to the lower corner of the opposite side). Find the size of each of these TV screens to the nearest inch.

a.

30 in.

40 in.

_____

b.

28 in.

36.5 in.

_____

c.

16 in.

21.75 in.

_____

4. In Figure 1 below, the ladder must be at least how tall to reach the top of the wall (to the nearest tenth of a foot)? _____

5. In Figure 2 below, the two cars left Oma at the same time. One car traveled due north at 60 miles per hour and the other traveled due east at 45 miles per hour. How far apart were the cars after 1 hour? _____

6. In Figure 3 below, Scruffy is directly beneath the kite. How high is the kite? (Be careful as to which side of the triangle you are looking for!) _____

Figure 1

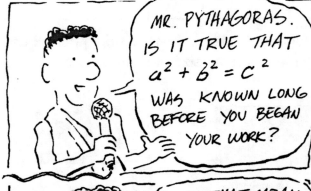

Wall 40 ft   Ladder (c ft)

10 ft

Figure 2

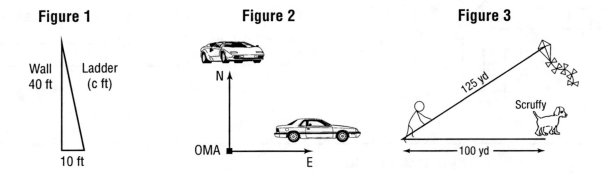

N

OMA

E

Figure 3

125 yd

Scruffy

100 yd

7. Find the distance from *A* to *B* to the nearest tenth of a unit. (Draw a right triangle on the grid with *AB* as one of its sides.)

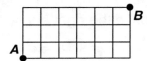

_____

Number Tricks

• •

- For each Number Trick, you are asked to pick a number and then perform a series of computations. Sometimes you are given the directions in the Verbal column; sometimes you are given the directions in the Algebra column.

- Complete the Verbal, Algebra, and Arithmetic columns for each Number Trick. Be sure to *simplify* the results in the Algebra column.

- Do the computations for the number that is picked in the Arithmetic column.

- After you complete the columns, write a generalization for the trick. A *generalization* is a conclusion that is true no matter what number is picked first.

Sample Number Trick

Verbal	Algebra	Arithmetic
a. Pick a number.	x	Let's pick 10.
b. Multiply by 2.	$2x$	2 x 10 = 20
c. Add 6.	$2x + 6$	20 + 6 = 26
d. Divide by 2.	$\dfrac{2x + 6}{2} = x + 3$	26 ÷ 2 = 13
e. Subtract the number you picked.	$x + 3 - x = 3$	13 − 10 = 3
f. Generalization: You will always end up with 3.		

Try out the Sample Number Trick: Ask someone to do steps **a–e** mentally. Then tell the person that his or her final result is 3 (no matter what number was picked first).

Number Trick 1

Verbal	Algebra	Arithmetic
a. Pick a number.	x	
b. Add 5.		
c.	$2(x + 5) = 2x + 10$	
d. Subtract 10.		
e. Generalization:		

Try out Number Trick 1: Have someone do steps **a–d** mentally. Then ask the person to tell you the number that he or she picked. If you double it, you can tell the person the final result for step **d.**

Number Trick 2

Verbal	Algebra	Arithmetic
a. Pick a number.	x	
b. Multiply by 4.		
c. Subtract the number you picked.		
d.	$3x + 12$	
e. Multiply by $\frac{1}{3}$.		
f. Generalization:		
g. Describe how you could try out Number Trick 2.		

Number Trick 3

Verbal	Algebra	Arithmetic
a. Pick a number.	x	
b. Add the number you picked.		
c.	$2x - 2$	
d.	$2(2x - 2) = 4x - 4$	
e. Multiply by $\frac{1}{4}$.		
f. Subtract the number you picked.		
g. Generalization:		

Name: _____   Date: _____

Planely Algebra

• •

United Airlines Flights Departing from Chicago		
Flight	**Gate-to-Gate Minutes***	**Miles**
to Boston	129	867
to Cleveland	65	316
to Dallas	138	802
to Denver	150	901
to Detroit	68	235
to Indianapolis	51	177
to Nashville	84	409
to New Orleans	125	837
to New York City	119	733
to Orlando	156	1005
to Toronto	85	437
to Washington, DC	103	612

*This is the combined time a plane spends on the runways taxiing and the time it spends in flight.

1. Graph the data points from the table above. Use the grid on the next page. Graph the "Gate-to-Gate Minutes" as the *x* values; graph the "Miles" as the *y* values. Scale the *x*-axis in intervals of 10 minutes. Do not connect the points on the graph. The graph you are making is called a *scatterplot.*

2. Notice that the data points are close to being on a line. We can *fit a line to the data* by using a ruler to draw a line that comes as close as possible to as many dots as possible. Use a ruler to draw such a line. Extend the line so that it crosses the *x*-axis. _____

3. You can use your line to help make some predictions for flights not given in the table.
   a. Estimate the miles for a flight that takes 170 gate-to-gate minutes. _____
   b. Estimate the gate-to-gate minutes for a 500-mile flight. _____

4. **a.** Estimate how much change there is in the height of your line (along the *y*-axis) for every change of 10 minutes (along the *x*-axis). This will give you the approximate distance a plane travels in 10 minutes. _____

b. Divide your result in Question 4a by 10 to find the speed of a
plane in miles per minute. This result, the change in the height
of a line for every change of 1 unit to the right, is called the
slope of the line.   _____

5. a. What is the ordered pair for the point where your line crosses the
x-axis? The *x*-coordinate of that point is called the *x-intercept*.   _____

b. What are the airplanes doing during the time represented by
the *x*-intercept?   _____

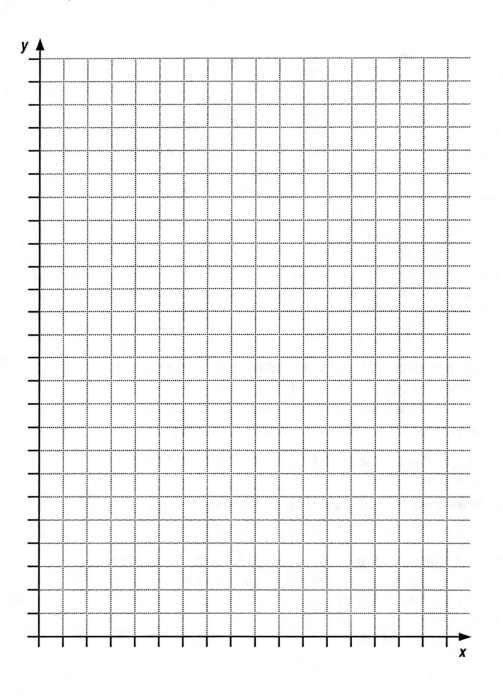

Upside-Down Calculator Problems

● ●

Use a calculator to do the mathematical portion of each problem. Then turn the calculator 180 degrees and read the "word" displayed. Use the words to fill in the blanks.

1. Anne and her four friends equally share the cost of a $28.02 pizza. While Anne computed each person's share of the cost, her friends ate up the pizza. When Anne realized that all of the pizza was gone, she said, "You guys are a bunch of _____ !"

2. Josie "I'm Awesome" Wosie was not allowed to play basketball because his grade-point average was only 0.1265307. However, the next semester he studied and obtained an average of 3.908148. The coach computed the amount of improvement and said, "Josie, you're now _____ to play."

3. Pick-Pocket Pete pilfered $500 from his partner Pat Powers. Pete escaped, puttering at a speed of 40 miles per hour for 8.777 hours. However, Pete then got a flat tire going over something pointy, and the police, computing how far Pete went, picked him up in the town of _____ .

4. Study how you can use an ordinary calculator to add fractions—without changing the fractions to their decimal equivalents!

   Use your calculator to find this sum: $\frac{3}{5} + \frac{8}{3}$

$$\frac{7}{5} + \frac{3}{4} \longrightarrow 7005 \times 3004 = 210\mathbf{43}0\mathbf{20}$$

$$\text{So, } \frac{7}{5} + \frac{3}{4} = \frac{43}{20}.$$

The *four digits* from your calculator display that are used to determine the sum will "spell out" what your reaction to this method might be: _____

5. A dealer bought a car for $11,050 and sold it at a 70% loss. After he computed how much money he lost on the deal, he decided not to _____ cars again.

6. Use the clues and indicated computations to complete this crossword puzzle. (Write the "words" in the crossword puzzle.)

Across

1. turkey talk
   $(1,000,000 - 52,977.5) \div 2.5 - 3$

4. Dutch painter
   $10 \times 20 \times (4 \times 5 + 3) + 6$

5. water cylinder
   $161 \div 4.6 + 0.04$

Down

1. Jack Benny made you_____.
   $8 \times (27,000 + 20,077)$

2. captain of the HMS Bounty
   $(2^3 \times 527) - 0.01 - 375,422$

3. female name
   $35,000 - 8.4 \times 2.5 + 2 \times 97$

From *Math for Real Kids* © 2005 David B. Spangler.

Activities

· ·

Know Your Bowling Score!   Use after pages 1–2.

Use a pair of dice to play a game of tabletop bowling.

- For the *first throw* in a frame, roll *two dice.* The number of pins you knock down is the sum of the numbers shown on the dice. If you roll a 10, 11, or 12, you have a *strike.*
- If you do NOT get a strike, roll *ONE die for your second throw.* Add the numbers from the two throws to determine the number of pins knocked down for the frame. If the number you roll on the second throw <u>is greater than or equal to</u> the number of pins still up (from your first throw), you have a *spare.*

Keep score of your games below. You may want to bowl with another person.

Name _____

1	2	3	4	5	6	7	8	9	10

Name _____

1	2	3	4	5	6	7	8	9	10

Presidential Playing Field   Use after pages 5–6.

You might enjoy playing out a "presidential election" with a partner. Make up a game that has 51 rounds—one for each state, plus Washington, D.C. The game could be as simple as guessing a coin flip for each round. It might be something athletic, such as shooting baskets or playing Ping Pong.

Start with, say, Alaska (AK). Because Alaska has 3 electoral votes, whoever wins the round gets all 3 of Alaska's electoral votes. Continue playing each state, with the winner of the round getting all of the electoral votes for the state being played. Whoever gets a majority of the 538 electoral votes wins the "election."

Formula-Driven

Use after pages 15–16.

Make up your own formula, such as the one in Question 9 on page 16. Provide some data for the formula, and have someone find the results.

Take Your Cue from Coupons

Use after pages 20–21.

Clip grocery coupons from a newspaper. Go to a store and find out how much money the items would cost three ways: (1) without coupons, (2) with only manufacturers' coupons, and (3) with both manufacturers' and any store coupons.

Problems That Are Just About Average

Use after pages 35–36.

Select topics for which you would like to collect and analyze data from your class. You may choose some of these topics or select some of your own.

- favorite ice-cream flavor
- eye color
- letters in name
- number of people in family
- number of pets
- hours spent per day watching TV
- favorite TV program
- favorite CD
- distance from home to school, in blocks
- age in months
- birthday month

- languages spoken at home
- favorite sport/game
- favorite expression
- instrument(s) played
- favorite meal/food
- rating (on a scale of 1 to 5) for a particular movie (or other event)
- yes-no response to a question, such as, "Do you know how to swim?"
- amount of money spent per week on entertainment, such as on video games

Write a questionnaire and collect the data. Then make a frequency table. Decide which measure (or measures) of central tendency to use to describe each kind of data. For example, you may find that the typical student favors chocolate ice cream, has brown eyes, lives with a family of 4, is 123 months old, and knows how to swim.

Pool the information into a Class Profile showing the characteristics of a typical member of the class. You could also make a drawing. Give your "composite student" a name by using letters that occur most frequently in the names of the students in your class.

Internet Extension: As an extension, collect data from students in another class. Use the Internet to seek out data from students in other cities, states, or countries! This could lead to some interesting comparisons.

Displaying Your Data: You might want to publish your results in a school newspaper or have a math fair and display your "composite student." Serve the favorite food and play the favorite music of your composite student.

From *Math for Real Kids* © 2005 David B. Spangler.

Take Stock of Stocks

Use after page 39.

How would you like to try to make some (pretend) money in the stock market? Begin with an account of $10,000. Over the next 8 weeks, use the money to buy and sell stocks of your choice. Stock prices are given in the business section of newspapers. Each time you buy or sell a stock, pay a commission of $50 from your account. At the end of 8 weeks, see who has made the most money.

Odds Are...

Use after pages 47–48.

Play a video game (or some other game), say, 10 times. Keep track of how many times you win or lose. Then compute the following: (a) your odds for winning; (b) your probability for winning; (c) your odds for losing; and (d) your probability for losing.

What Shape Is He In?

Use after pages 51–52.

Write a story of your own that can be acted out with tangrams.

What's Cooking?

Use after page 54.

Choose a recipe. Increase the recipe's yield, as needed, to make enough for your class. Find out how much the ingredients cost. Determine the unit cost (cost per slice, per piece, etc.) if you were to make and sell the food. Write a marketing slogan or campaign that could be used to help "sell" your product.

A Weighty Matter

Use after pages 65–66.

Find packages of food or other objects that give measurements in both metric and customary units of length, capacity, or mass. On each container, cover one of the measurements with masking tape. Bring the containers to school and see if your classmates can guess the covered measurements.

Try These—to a Degree

Use after pages 67–68.

Use a newspaper to find data on the daily high and low temperatures for a city of your choice. Make a graph to display the data for a week and add some drawings to illustrate the weather. Write questions based on the graph for classmates to answer. If a camcorder is available, write a script for your weather data and tape your own weather report.

In this activity you will be making a self-portrait.

Materials: • translucent marker paper (or tracing paper)
• soft lead pencils and colored pencils
• a mirror

Procedure:

1. Place marker paper over the grid on the following page, leaving several inches more space at the top of the grid than at the bottom to allow space for drawing the hair. First, examine the shape of your face by observing yourself in the mirror. Then draw the shape of your face in rectangle *ABGK*. Be sure that your face fills the entire rectangle.

2. Draw the eye openings, which are almond shaped, so that line *D* runs through them. The inside corners of the eye openings (nearest the nose) should touch where line *D* intersects lines *J* and *H*. The eyelid partially covers the iris, so don't draw a complete circle for the iris. Two-thirds of the iris should extend above line *D*. The pupils should be resting above line *D*.

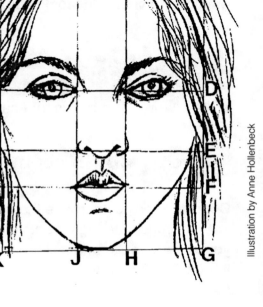

3. Draw the lower part of the nose on line *E*. Outline the nostrils and the shapes around them. Take care to draw the shapes accurately. Don't draw a vertical outline around the nose. You may use shadowing later in order to give dimension.

4. Draw the lips around line *F*. First draw the line that parts the lips. Next, draw the upper and lower lips.

5. The ears extend from above line *D* to about line *E*. The earlobes will fall below *E*.

6. Draw in your hair, which may go beyond the grid in the extra space allotted.
   Note: You are drawing a mirror image of your face. So, if you part your hair on the left side, it will appear on the right side. To rectify this, flip the drawing over and either place another sheet over the drawing and trace, or just darken the lines on the opposite side of the paper.

7. After drawing in the features, color them by blending the colors using the broad side of the exposed lead rather than the point of the colored pencils.

From *Math for Real Kids* © 2005 David B. Spangler.

Grid for Self-Portrait

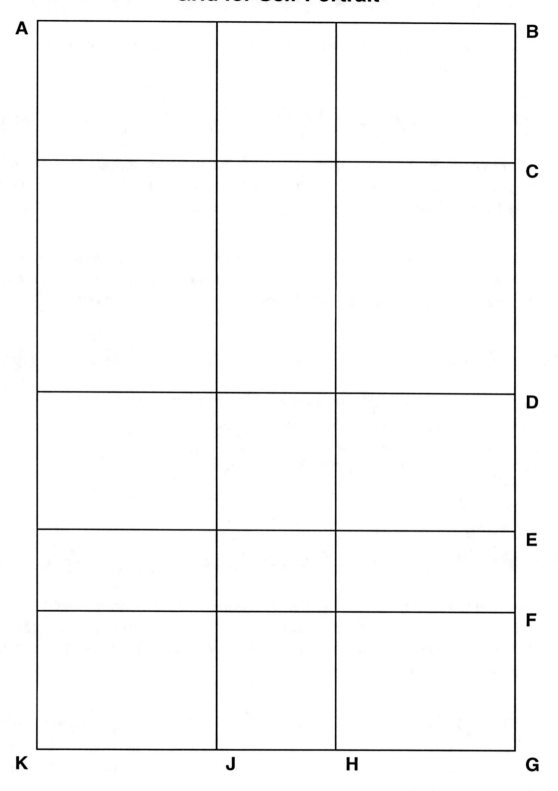

When You Finish: Look for golden ratios in the grid. For example, consider line segment *BD* and its section *CD*. Note that $\frac{BD}{CD} \approx 1.6$. Find more such golden ratios. (See page 116 for answers.)

Spread Out Your Spreadsheets

Use after pages 82–83.

Modify your spreadsheet to change the amount invested, interest rate, and number of years. Make up and solve your own problems.

Turn Up the Volume

Use after pages 92–93.

Use grid paper to help you design and make a model of your own cereal box. Be sure your model includes all six faces. (The grid shown in Question 1 on page 92 makes an "open" box with five faces.) Try to make a box with the largest possible volume from the grid paper you have.

Give your cereal a name and draw some art on the box. Include other information on the box, such as the weight of the cereal, price, unit price, nutritional data, and volume of the box. You may need to refer to real cereal boxes to help you estimate some of the data.

Hidden Figures

Use after pages 96–97.

Create your own "Hidden Figures" problem for someone to solve. You do not have to limit the hidden figures to squares or triangles. For example, you may want to have people look for parallelograms or trapezoids in a figure like the one that is shown at the right.

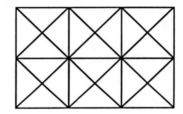

Getting to the Root of the Problem

Use after page 102.

Design a flowchart to describe a procedure of your choice. The procedure might be for making a phone call, making an egg, or getting up in the morning. You might choose a mathematical procedure, such as adding two fractions. Include at least one decision that must be made based on a yes-no question.

Number Tricks

Use after pages 105–106.

Make up your own Number Trick that has at least four steps in the Verbal, Algebra, and Arithmetic columns. Try it out on someone.

This Ratio is Golden, Answers: $\dfrac{DG}{DF}, \dfrac{DG}{EG}, \dfrac{GK}{GJ}, \dfrac{GK}{HK}$; rectangle *ABGK* is golden.

Answer Key

Know Your Bowling Score!, pages 1–2

1.

1	2	3	4	5	6	7	8	9	10
6 3	8 -	5 /	4 5	6 /	9 -	- /	2 6	7 /	5 4
9	17	31	**40**	**59**	**68**	**80**	**88**	**103**	**112**

2.

1	2	3	4	5	6	7	8	9	10
9 -	X	7 2	6 /	9 -	X	5 4	- /	X	5 / 9
9	28	**37**	**56**	**65**	**84**	**93**	**113**	**133**	**152**

3.

1	2	3	4	5	6	7	8	9	10
8 -	X	X	4 5	6 /	X	X	3 5	1 /	X 3 4
8	32	51	**60**	**80**	**103**	**121**	**129**	**149**	**166**

4.

1	2	3	4	5	6	7	8	9	10
2 /	9 -	X	X	7 /	- 9	5 /	X	X	X 5 2
19	**28**	**55**	**75**	**85**	**94**	**114**	**144**	**169**	**186**

5.

1	2	3	4	5	6	7	8	9	10
X	X	X	6 3	X	- /	5 4	X	X	X X X
30	**56**	**75**	**84**	**104**	**119**	**128**	**158**	**188**	**218**

6. 300

Counting Calories, pages 3–4

1. 4,350 calories

2. *Charlie Calorie, age 12, weight 98 lb*

Food	Calories
Breakfast	
doughnut, iced (1)	210
pancakes, 3 with syrup	360
milk, whole (1 cup)	166
orange juice (1 cup)	110
Lunch	
hamburger (1/4 lb)	334
potatoes, fried (20)	394
milk shake (10 oz)	355
candy, milk chocolate (1 oz)	145
Dinner	
chicken, fried (2 drumsticks)	390
salad and dressing	119
corn, cooked (1 ear)	85
cola (12 oz)	160
pie, apple (1 slice)	377
popcorn, air-popped (1 cup)	30
TOTAL	**3,235**

3. *Nancy Nutrition, age 13, weight 100 lb*

Food	Calories
Breakfast	
cornflakes, unsweetened (1.1 oz)	110
egg, fried (1)	110
milk, 2% (1 cup)	130
banana (1)	104
orange juice (1 cup)	110
Lunch	
bread, white (2 slices)	126
peanut butter (1 tablespoon)	95
jam (1 tablespoon)	55
apple (1)	100
milk, 2% (1 cup)	130
Dinner	
turkey, roasted (5 oz)	240
potato, baked (1)	98
margarine (1 tablespoon)	100
corn, cooked (1 ear)	85
carrots, small (2)	50
milk, 2% (1 cup)	130
ice cream (1 cup)	270
TOTAL	**2,043**

4. Charlie Calorie's

Presidential Playing Field, pages 5–6

1. 16 electoral votes
2. 19 electoral votes
3. 30 electoral votes
4a. 189 electoral votes
4b. 349 electoral votes

5a. Cleveland
5b. 95,713 popular votes
5c. Harrison
5d. 65 electoral votes
5e. Harrison

6a. 1888
6b. 2000
7. 270 electoral votes
8. 59,516,760 popular votes

9. Yes
10. neither candidate

A Trip to 1776, pages 7–8

1. Answer depends upon current year: current year minus 1776.

2. 144 years
3. 10,000 troops
4. $100 bill
5. 342 chests

6. 1826
7a. 2,265 miles
7b. 1873
8. 93 years

9. 30 delegates
10a. June 14, 1877
10b. 37 more stars
10c. 141 years later

Multiply or divide.

1. 4 x 8 = 32
2. 28 ÷ 4 = 7
3. 0 x 9 = 0
4. 7 x 8 = ~~54~~ *56*

5. 36 ÷ 9 = 4
6. 1 x 12 = ~~1~~ *12*
7. 12 ÷ 3 = ~~36~~ *4* or 12 x 3 = 36
8. 8 ~~x~~ 6 = 48

9. 42 ÷ 7 = 6
10. 90 ÷ 9 = ~~9~~ *10*
11. 11 ÷ 1 = 11
12. 9 x 9 ~~=~~ 81

13. 10 x 11 = ~~101~~ *110*
14. 45 ÷ 5 = 9
15. 3 x 3 x 0 = ~~9~~ *0*
~~15.~~ *16.* 0 ÷ 4 = 0

Solve each problem.

~~17.~~ *18.* Gertrude walks 4 miles each day. How many miles does she walk in 12 days? *insert space* — *miles* ~~48 days~~

~~18.~~ *17.* A baseball team gets 3 outs in an inning. How many outs does a team get in 9 innings? — *27 outs*

~~18.~~ *19.* Mrs. Smith's class has $96 to spend on class gifts. if each gift costs $8, How many gifts can the class buy? *≡ cap* — *12* ~~18~~ gifts

20. lower case

~~19.~~ *19.* A sheet of stamps has 10 rows with 10 stamps in each row. You tear off 2 full rows. How many stamps are left in the sheet? — *80* ~~8~~ stamps

~~20.~~ *21.* Start with the number 72. How many times would you have to subtract 6 to reach 0? *lower case* — *12 times* ~~66~~

~~21.~~ *22.* Frank Furter cooked 11 hot dogs. He cut each hot dog into 4 pieces. He needs 48 pieces for a party. How many more hot dogs must he cook? — *1* ~~4~~ hot dogs

A Wall-to-Wall Product, page 10

1a. Sample answer: Find the number of shaded tiles in one section.
1b. 15 tiles
1c. 120 tiles
2a. 392 tiles
2b. 272 tiles
3a. $1,960
3b. $2,200
4a. 1,568 tiles
4b. 3,528 tiles
4c. 16 times as many

A Bit of Computing, page 11

1. 8,000 bits
2. 1,024 bytes
3. 65,536 bytes
4. 1,048,576 bytes
5. 268,435,456 bytes
6. 1,073,741,824 bytes
7. 222,222,222
333,333,333
444,444,444
555,555,555

Camp Quo-Tent, pages 12–13

1. $147 per week
2a. 13 campers
2b. 14 campers
2c. 12 campers
3a. $100 per day
3b. $105 per day
4a. 468 campers
4b. 156 campers
5. 5 campers
6. 4 buses
7. 8 tents will have 5 people each; 5 tents will have 4 people each.
8. 10 teams
9. 1 letter per day
10a. 3 times as great
10b. 4 times as great
11a. 33 division problems
11b. 11 division problems

Stuff You Auto Know About Mileage, page 14

1a. 26,778

1b. 14 km per liter

2. 2,176 km

3. 10,482 km

4. 1,342 km

5a. 13 times

5b. 9 km

6. 112 liters

7. $157

Formula-Driven, pages 15–16

1a. 48°F

1b. 70°F

1c. 81°F

1d. 103°F

2. 37°F

3. 2 in. per sec

4a. 88°F

4b. 86°F

4c. 80°F

4d. 72°F

5. faster

6. 160 chirps per min

7. 37°F

8. 5 in. per sec

9. 56 loads

Can You Make the Change?, page 17

1. Suppose the *amount due* is $0.32, and *the amount given to you* is $1.

In making change, you give the customer 3 pennies and say, "$0.35."

Then you give the customer 1 nickel and say, "$0.40."

Then you give the customer __1__ dime and say, " __$0.50__ ."

Finally, you give the customer __2__ quarters and say," __$1.00__ ."

The amount of change is $0.03 + $0.05 + $0.10 + $0.50, or $ __$0.68__ .

2.

Amount due: **$1.18**

Amount given to you: **$2**

Change		What you say
2	pennies	$1.20
1	nickel(s)	$1.25
0	dime(s)	——
3	quarter(s)	$2.00

The amount of change is $ __$0.82__

	Amount Due	Amount Given to You	Change						Amount of Change
			$0.01	$0.05	$0.10	$0.25	$1.00	$5.00	
3.	$2.37	$5.00	3	0	1	2	2	0	**$2.63**
4.	$0.84	$1.00	1	1	1	0	0	0	**$0.16**
5.	$0.29	$1.00	1	0	2	2	0	0	**$0.71**
6.	$2.33	$3.00	2	1	1	2	0	0	**$0.67**
7.	$1.78	$5.00	2	0	2	0	3	0	**$3.22**
8.	$4.02	$10.00	3	0	2	3	0	1	**$5.98**
9.	$5.41	$10.00	4	1	0	2	4	0	**$4.59**
10.	$14.58	$20.03	0	0	2	1	0	1	**$5.45**

To Life!, pages 18–19

1a. 1920

1b. 1980

2a. 20.8 years

2b. 25.2 years

3a. 1940 and 1950; 4.8 years

3b. 1920 and 1930; 7 years

4. 1.6 years

5. Sample answer: The life expectancies for males and females have increased for each year in the table (1920–2000). The life expectancy for females has always been greater than it has been for males.

6. 57.5 years

7. 63.84 years; 71.34 years

8. Answers will vary. A male born in 1995, for example, could estimate that his life expectancy falls midway between 71.8 years and 74.4 years, or 73.1 years.

9. 24.0 years

10a. Rwanda 39.3, India 62.5, Egypt 63.3, China 71.4, Mexico 71.5, U.S. 77.1, Israel 78.6, Italy 79.0, Canada 79.4, Japan 80.7

10b. 41.4 years

Take Your Cue from Coupons, pages 20–21

1a. generic peanut butter
1b. $0.23 more
2a. $0.26 cheaper

2b. $0.49 cheaper
3a. $5.78
3b. $3.67

4a. $7.64
4b. $5.59
4c. $5.71

5. $3.98

Checking Your Checking Account, pages 22–23

1a. All items in the checkbook should be marked with a ✓ except check #641 to Josey Wosey, Inc., check #643 to Plug Outlet Store, the deposit on 4/1 for $100, and check #644 to Kent Reed.
1b. $100.00
1c. $76.15

1d. $1,270.69
1e. $1,170.69
1f. No. The $8.50 monthly checking fee needs to be entered into the checkbook

1g. $1,186.04
2. $1,156.26

More Power to You, page 24

1. 273
2a. 288 calories
5.

2b. 342 calories
3a. 191.25 calories

3b. 171.7 calories
3c. 132.6 calories

4. Yes

Appliance	Hours Turned On	Watt-hours	Kilowatt-hours (kWh)	Total Cost (nearest cent)
a. 100-watt light bulb	24.00	**2400.0**	**2.4000**	**$0.20**
b. 6,700-watt electric oven	1.50	**10,050.0**	**10.0500**	**$0.82**
c. 850-watt microwave oven	0.15	**127.5**	**0.1275**	**$0.01**
d. 225-watt TV	8.50	**1912.5**	**1.9125**	**$0.16**

At Any Rate . . . , page 25

1a. 560 miles per hour
1b. 51.9 miles per hour
1c. 0.06 mile per minute
2. 84 people per square mile

3. $80,000 per second ($0.08 million)
4a. 10 oz: $0.245 per ounce; 15 oz: $0.231 per ounce; 22 oz: $0.215 per ounce;

best buy: the 22-oz box
4b. 10 oz: $0.170; 15 oz: $0.181; 22 oz: $0.180; best buy: 10-oz box.

5. Highest yearly earnings: $13 per hour; lowest yearly earnings: $25,000 per year

Super Sports Statistics I: Baseball and Basketball, pages 26–27

1. Suzuki: .372
Ramirez: .308
Pujols: .331
Helton: .347
2. Sheets: 2.70
Zambrano: 2.75
Schilling: 3.26
Santana: 2.61

3. New York: .623
Boston: .605
Baltimore: .481
Tampa Bay: .435
4a. .552
4b. better

5a. .517
5b. .659
5c. better
6. McGrady 1,591
Stojakovic 1,862
Garnett 2,717

Super Sports Statistics II: Football, pages 28–29

1.

Quarterback	Result from Step 1	Result from Step 2	Result from Step 3	Result from Step 4	NFL Passer Rating
a. Steve McNair	1.625	1.259	1.200	1.938	100.4
b. Peyton Manning	1.848	1.135	1.025	1.933	99.0
c. Daunte Culpepper	1.749	1.166	1.101	1.769	96.4

Wheels and Deals, page 30

1a. 341.9 miles
1b. 26.5 miles per gallon
2. 345.6 miles
3a. $1,494

3b. $498
4a. 17.5 meters
4b. 27.5 meters

5a. $22,399; $5,600 (This answer could be $5,599 if the mark-up is computed *before* rounding.)

5b. $16,737; $3,013
5c. $14,875; $1,785

Learning About the Mean, page 31

1. 78
2. 912,500
3. 48.9
4. c

5. True
6. Yes. Each score in the set could be replaced by the mean without changing the sum.

Thus, you could multiply the mean by the number of scores to obtain the sum of the scores.

Learning About the Median, page 32

1. 87
2. 275,000
3. 40.5
4. b

Learning About the Mode, page 33

1. 85
2. The mode does not exist.
3. 77 and 91
4. Blue and Brown
5. b and d

Finding Averages from a Frequency Table, page 34

1. Number of people in each car	Tally	Frequency
1	₩₩₩ ₩₩₩ ₩₩₩ ₩₩₩ III	23
2	₩₩₩ II	7
3	₩₩₩ II	7
4	₩₩₩	5
5	IIII	4
6	III	3
7	II	2

2. 1
3a. 51 scores
3b. 2
4. The mean is 2.5 people per car.

Score	f	Score x Frequency
1	23	23
2	7	14
3	7	21
4	5	20
5	4	20
6	3	18
7	2	14
	51	130

Problems That Are Just About Average, pages 35–36

1a. Mean: 84; Median: 92; Mode; 99

2a. the median

2b. Sample answer: The mode is too high. (Only one score is above it.) The mean is too low. (Only two scores are below it.) The median, of course, is right in the middle.

3. Mean: 93; Median: 93; Mode: 99

4. Mean: $50,000; Median: $28,000; Mode: $20,000

5. the mode

6. Mean: $29,000; Median: $24,000; Mode: $20,000

7. the mean

8. the mode

9. the median

10. 555

11a. 99

11b. Sample answer: Five scores with a mean of 88 would be a total of 5 x 88, or 440 points. You have a total of 341 points. So you must score at least 440-341, or 99, on the next test.

12. Mean: 87; Median: 85; Mode: 80

13.

Course	Credits	Grade	Honor Points for Course
Math	1	A	4
English	1	B	3
Science	1	A	4
History	1	C	2
Art	0.5	B	1.5
P.E.	0.25	B	0.75
Totals	4.75		15.25

Grade-point Average 3.21

Where Do You Draw the Line?, page 37

1.

Year	Job A Earnings During Year	Job B Earnings During Year
1	$26,000	$26,800
2	$28,000	$28,400
3	$30,000	$30,000
4	$32,000	$31,600

2. See graph.

3. Job B

4. Year 3

5. See graph.

6. $1,200

7. $38,000

8. increase

9. Job B

2. & 5.

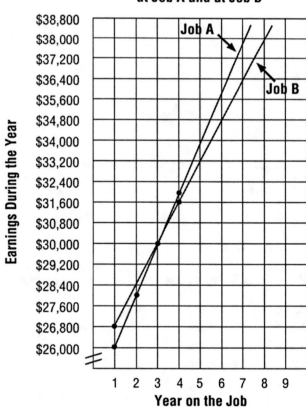

Earnings During Each Year at Job A and at Job B

Graphs That Should Be Barred, page 38

1a. Something *has* been "eating away" at the profits. The profits have been *decreasing* since 2002—not increasing.

1b. The years have been positioned in *reverse order.* They should read from 2002 to 2004—not from 2004 to 2002.

2a. 5 times as tall

2b. The scale on the vertical axis does not begin at 0, so the reader is only looking at the *top* of the graph. This magnifies the difference in the number of votes

Chris received compared to those received by the other candidates.

3a. 3 times as many people

3b. (1) The vertical scale is *not uniform.* It initially increases by 10, then it jumps by 30. (2) The bar for Brand C is *twice* as wide as the other bars. Each bar should be of uniform width.

3c. The changes make Brand C look much better, relatively speaking, than it should.

4.

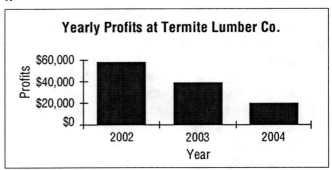

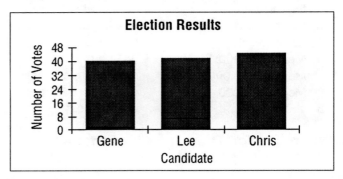

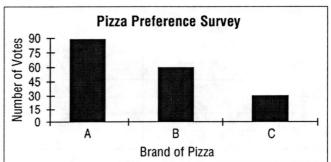

Take Stock of Stocks, page 39

1. $\frac{1}{2}$ dollar

2. $\frac{3}{4}$ dollar

3. $1\frac{3}{8}$ dollars

4. $24\frac{1}{2}$ dollars

5. $2\frac{1}{4}$ dollars

6. $25

7. $63

8. $21

9. $8\frac{1}{4}$ dollars

What Do They Have in Common?, pages 40–41

1a. every 18 squares
1b. every 36 squares
1c. every 60 squares
1d. every 36 squares
2. 30 days
3. 2089
4a. 56
4b. 36
4c. 60

4d. 168
5. March 31
6a. 24 pens
6b. 25 pens
6c. Brand Y
6d. The LCM of 6 and 9 is 18. With Brand Y, you get 15 pens for $18. With Brand Z,

you get 14 pens for $18. So with Brand Y, you get more pens for the same money.

7. The LCM of a set of numbers is the largest number in the set when the largest number is a multiple

of each of the others.

8. The LCM of a set of numbers is the product of all the numbers in the set when the greatest common factor of all of numbers is 1.

Rulers Rule!, page 42

1a. $\frac{5}{8}$ in.

1b. $\frac{5}{16}$ in.

1c. $8\frac{1}{2}$ in.

1d. $16\frac{3}{4}$ in.

1e. $18\frac{11}{16}$ in.

1f. $9\frac{7}{8}$ in.

2a. $\frac{15}{16}$ in.

2b. $17\frac{3}{8}$ in.

2c. $25\frac{1}{4}$ in.

2d. $27\frac{3}{16}$ in.

2e. $35\frac{7}{16}$ in.

2f. $18\frac{11}{16}$ in.

3a. $40\frac{5}{6}$ yd

3b. $31\frac{19}{20}$ mi

3c. $65\frac{31}{40}$ mi

4. $7\frac{1}{4}$ in.

Mixed Numbers at the Olympics, pages 43–44

1a. 9 ft $2\frac{1}{4}$ in.

1b. 14 ft $\frac{1}{2}$ in.

1c. 1 ft 11 in.

1d. Less than 4 inches. Because $\frac{1}{3} > \frac{1}{4}$, the difference between $8\frac{1}{4}$ in. and $4\frac{1}{3}$ in. is less than 4 inches.

2a. 1 ft $5\frac{1}{2}$ in.

2b. 10 ft $\frac{3}{4}$ in.

2c. 1992; 6 in. better

3a. 4 ft $11\frac{3}{4}$ in.

3b. Answer is dependent upon the current year: 656 + (current year minus 1).

3c. 29 ft $2\frac{1}{2}$ in.

3d. 28 ft $\frac{1}{4}$ in. (1984) and 27 ft $10\frac{3}{4}$ in. (1996)

3e. $1\frac{1}{2}$ in.

3f. 2 in.

4a. 245 ft

4b. $74\frac{33}{48}$ ft

4c. $226\frac{1}{12}$ ft

5. 4 min $31\frac{981}{1000}$ sec

Spinning into Probability, page 45

1a. 5

1b. Yes. The size of each sector is the same.

2a. 1

2b. 1 chance out of 5

2c. $\frac{1}{5}$

2d. 0.2

3a. 2

3b. 2 chances out of 5

3c. $\frac{2}{5}$

3d. 0.45

4a. 5 chances out of 5

4b. $\frac{5}{5}$

4c. 1

a. 0 chances out of 5

b. $\frac{0}{5}$

5c. 0

6a. 1

6b. 0

6c. 0 and 1

Chance Problems, page 46

1a. 8

1b. $\frac{1}{4}$

1c. $\frac{3}{4}$

1d. $\frac{1}{8}$

1e. 0

1f. 1

1g. $\frac{1}{2}$

2a. $\frac{1}{4}$

2b. $\frac{3}{4}$

3a. $\frac{1}{3}$

3b. $\frac{2}{3}$

4a. $\frac{1}{36}$

4b. $\frac{2}{36}$ or $\frac{1}{18}$

4c. $\frac{3}{36}$ or $\frac{1}{12}$

4d. $\frac{4}{36}$ or $\frac{1}{9}$

4e. $\frac{5}{36}$

4f. $\frac{6}{36}$ or $\frac{1}{6}$

4g. $\frac{5}{36}$

4h. $\frac{1}{36}$

4i. 0

5. $\frac{4}{36}$ or $\frac{1}{9}$

Odds Are . . . , pages 47–48

1a. $\frac{1}{2}$ **1d.** $\frac{2}{3}$ **2c.** 1 to 2   **3b.** 10 to 3

1b. $\frac{1}{2}$ **2a.** 1 to 1   **2d.** 2 to 1   **4a.** $\frac{1000}{1003}$

1c. $\frac{1}{3}$ **2b.** 1 to 1   **3a.** 3 to 10   **4b.** $\frac{3}{1003}$

5.

Team	Odds For Winning	Probability For Winning	Odds Against Winning	Probability Against Winning
a. Bears	5 to 2	$\frac{5}{7}$	2 to 5	$\frac{2}{7}$
b. Packers	**5 to 7**	$\frac{5}{12}$	**7 to 5**	$\frac{7}{12}$
c. Lions	1 to 1	$\frac{1}{2}$	**1 to 1**	$\frac{1}{2}$
d. Vikings	**1 to 2**	$\frac{1}{3}$	2 to 1	$\frac{2}{3}$

6. The corresponding probabilities do not "add up." The sum is $\frac{1}{3} + \frac{1}{2} + \frac{2}{3}$, or $1\frac{1}{2}$. However, since the sum of the probabilities of all possible non-overlapping events in a sample space must equal 1, the spokesperson's calculations are in error.

It's Predictable, pages 49–50

1a. 15 times

1b. 30 times

2a., b.

	Land	Ware	Aska	Ana
Probability	$\frac{21}{50}$	$\frac{15}{50}$ or $\frac{3}{10}$	$\frac{13}{50}$	$\frac{1}{50}$
Prediction	315	225	195	15

3a. $\frac{1}{5}$ **4a.** 4 points   **4c.** 0 points   **6.** 25,000 incorrect prescriptions

3b. $\frac{4}{5}$ **4b.** 4 points   **5.** 68 people

Tangram #1	Tangram #2	Tangram #3
Tangram #4	Tangram #5	Tangram #6
Tangram #7	Tangram #8	Tangram #9

Tangrams, Fractions, and Area, page 53

1.

Tan	A	B	C	D	E	F	G
Area (in square units)	8	8	2	4	2	4	4

2. 32 square units

3.

Tan	A	B	C	D	E	F	G
Area (in square units)	2	2	$\frac{1}{2}$	1	$\frac{1}{2}$	1	1

4. Sample Solution:

- The "fold" along the diagonal divides the large square into two congruent parts. Thus, *A* and *B* together comprise $\frac{1}{2}$ the area of the large square. Since *A* and *B* are congruent, each is $\frac{1}{2}$ X $\frac{1}{2}$, or $\frac{1}{4}$, of the whole.
- *C, D,* and *E* can be placed on top of *A* to show that the sum of their areas is equal to that of *A*, or $\frac{1}{4}$, of the whole. Also, *C* and *E* fit perfectly on *D*. Hence, four pieces the size of *E* (or *C*) are equal in area to that of *A*. So, *E* and *C* are each $\frac{1}{4}$ X $\frac{1}{4}$, or $\frac{1}{16}$, of the whole.
- Since *D* is twice the size of either *E* or *C*, *D* is 2 X $\frac{1}{16}$, or $\frac{1}{8}$, of the whole.
- Since *C* and *E* can be placed on *F* to match perfectly, *F* is also 2 x $\frac{1}{16}$, or $\frac{1}{8}$, of the whole.

What's Cooking?, page 54

1a. $\frac{3}{8}$

1b. $\frac{3}{4}$

1c. $9\frac{3}{4}$

1d. $1\frac{1}{2}$

1e. $4\frac{1}{2}$

1f. $15\frac{3}{4}$

1g. 24

2. 9

3a. 3 hr

3b. $3\frac{1}{2}$ hr

3c. $4\frac{5}{16}$ hr

3d. $4\frac{11}{16}$ hr

4. 8 cups

5. $3\frac{1}{2}$ cups

6. $6\frac{7}{8}$ cups

It's Out of This World, pages 55–56

1a. $16\frac{5}{12}$ days

1b. $79\frac{4}{5}$ times as long

2a. about 6 months

2b. 16 times

2c. $1\frac{1}{2}$ hours

3. about $\frac{2}{5}$ million, or or 400,000 miles per day

4. 10 km per sec

5. 18 million km per min

6. 4 light-years

7. 40,000,000 times as long

8. 153 pounds

9. $123\frac{3}{4}$ pounds

10. about 1,975 miles

The Spy Who Mixed Numbers, pages 57–58

1. $13\frac{1}{3}$ yd

2. 20 yd

3. 8 yd per sec

4. $1\frac{2}{3}$ hours

5a. $g = 20\frac{1}{2}$

5b. $h = 30\frac{3}{4}$

5c. $i = 51\frac{1}{4}$

6. 2 cookies

7. $20\frac{1}{2}$ cans of cola

8. 1,681 bags of chips

In My Estimation . . . , pages 59–60

1. b

2. c

3. b

4a. b

4b. No

4c. Answers may vary. If you enter 6609 into the calculator instead of 609, you will obtain the incorrect sum, 9338.

5. c

6. c

Answers will vary for the estimates in **7–12.** It is suggested that you accept any *reasonable* estimate.

7. about 650 feet

8. more than 7 gallons

9a. about 59,000,000

9b. about 3 times as great (or slightly more than 3 times as great)

10. about 70 inches

11a. around .45

11b. perhaps .90 to .95

12a. Monthly payments of $1, $2, $4, $8, and so on, have the greater value.

12b. About $30,500 more

13. Answers will vary. There are about 9,000 hours in a year. So a 12-year-old has lived about 108,000 hours. (Think of 24 hours in a day as being about $\frac{1}{4}$ of 100. If you think of a year as having about 360 days, then think of $\frac{1}{4}$ of 360 = 90. So the product 24 x 365 is equal to about 9,000.)

14. over 100 years

Monster Math, page 61

Answers on this page may vary due to differences in measurements.

1a. 6 cm

1b. 480 km

2a. 3.5 cm

2b. 280 km

3a. 9.7 cm

3b. 776 km

4. Dracula's Coffin

5a. 2.5 cm

5b. 1400 km

6. 2744 km

7a. 7.2 cm

7b. 4032 km

Going to Great Lengths, pages 62–63

1. mm

2. m

3. cm

4. m

5. km

6. mm

7. km

8. cm

9. 1 meter

10. 1 mile

11. 50 km

12. 4.5 mm, or 0.45 cm

13. 20 lengths

14a. 3.9 km

14b. 2.9 km

15. 1.661 m

From *Math for Real Kids* © 2005 David B. Spangler.

In What Capacity?, page 64

1. L
2. mL
3. L
4. mL
5. kL
6. 500 mL
7a. an 11-minute shower
7b. 38.325 kL
L8. 5 days

A Weighty Matter, pages 65–66

1. kg
2. mg
3. t
4. kg
5. g
6. t
7a. 150 mg
7b. 0.25 g
8a. 882 newtons
8b. 100 kilograms
8c. 1.666 newtons
9. 39 kg
9b. 117 kg
9c. 98 kg

Try These—to a Degree, pages 67–68

1. 35°C
2. 100
3. 0
4. 37
5. 25
6. −10
7. 30
8. 15
9. 200
10a. 32
10b. 30
10c. 68
10d. 37
10e. 2
10f. 3632
11a. 34
11b. 130
11c. 23
11d. 4
12a. 60 degrees
12b. Sunday, Monday, and Friday
12c. 37°C
(Answers may vary slightly depending upon estimates made from graph.)

It's About Time, page 69

1a. 3 times
1b. 23 times
2. 3 hr 17 min
3. 4 hr 10 min
4. 53 min
5. 42 mi/h
6a. 2.25 hours on Wednesday; 7.75 hours on Saturday. Total: 13.5 hours
6b. $127.44
7. 19 times

This Ratio Is Golden, page 70

Answers may vary slightly due to differences in measurement readings.

1a. length: 3.2 cm; width: 2.0 cm; ratio of length to width: 1.6
1 b. length: 3.5 cm; width: 1.4 cm; ratio of length to width: 2.5
1c. length: 2.4 cm; width: 1.5 cm; ratio of length to width: 1.6
1d. length: 2.7 cm; width: 2.2 cm; ratio of length to width: 1.2
1e. Answers will vary. Many people would view the rectangles in **a** and **c** to be the most pleasing (because they are golden).
2. The rectangles in **a** and **c** are golden.
3a. ratio of length to width: 1.6; Yes
3b. ratio of length to width: 1.67; Yes
3c. ratio of length to width: 1.29; No

A Portion of Proportions, pages 71–72

1a. 7 times
1b. 12 times
1c. 36 times
2a. 35 times
2b. 15 times
2c. Jamie's; 28 more times
3a. 1,800 votes
3b. 14,525 votes
4. 69.12 meters
5. 17 minutes
6a. $1.81
6b. $1.73
6c. $1.49
7a. 8 feet
7b. 120 pounds

Percent Potpourri, pages 73–74

1a. Able and Mable
1b. 66 more squares
2a. $1.12
2b. $2.08

3. $420,000
4. 19,026 innings
5. 24%
6a. 300 girls

6b. 705 boys
6c. 33% (1005 teenagers out of 3000 is roughly 33%.)

7. 255 boys

Building a Solid Foundation on Percents, pages 75–76

1. 57.1%
2a. 9.4%
2b. 2.3%
2c. 12.4%
3a. the Chrysler Building
3b. 32.1%

4. 1,776 feet
5a. the Empire State Building
5b. Answer depends upon the current year. Subtract 1913 from the current year.

Divide 41 by that difference.
6. 75 grams
7. 25%
8. $2,399
9. 100%

10a. 1.7%
10b. 89.3%
11. 7,200 people

Percents That Don't Make Sense, page 77

1a. It would be free.
1b. 71.4%
1c. $85.65. (The first 100% would make the stapler free. The remaining 614% of the list price would be "paid to you." 614% of $13.95 = $85.65.
2a. No
2b. They computed "males as a percent of females" rather than "males as a percent of the total."

2c. 1990: 45%; 2000: 44%
3a. Answers may vary. Two of the percent increases should be 100% or more.
3b. Village Green: 35%; ClubGolf:100%; GolfClub: 233%
3c. The 1996 cost was divided by the 1992 cost. The result was converted to a percent by moving the decimal point one

place to the right. (Actually, the correct answers could be obtained by dividing the 1996 cost by the 1992 cost. In the case of Village Green, the quotient, 1.35, is obtained. The correct interpretation of this is that it cost *135% more* in 1996 than it did in 1992. This translates to a 35% increase. For

ClubGolf, the quotient is 2. This means it cost *200% more* in 1996. This translates to a 100% increase.)

Take Interest in This, pages 78–79

1. $36.88
2. $3,703.35

3. $2,976.94 (or $2,976.92, depending upon rounding)

4. $675
5. $9,207.50
6. $11,262.13

7. 24%
8. 22%

Take Even More Interest in This, pages 80–81

1a. $56.10
1b. $1,056.10
1c. $59.25
1d. $1,115.35

3a. 12 years
3b. 6 years
3c. 6.7 years

3d. 20.3 years
4a. $27,968.24
4b. $13.50

2.

Year	Principal	Interest	Amount in Account
1	$2,500.00	$197.50	$2,697.50
2	$2,697.50	$213.10	$2,910.60
3	$2,910.60	$229.94	$3,140.54
4	$3,140.54	$248.10	$3,388.64

Spread Out Your Spreadsheets, pages 82–83

Questions 1–11 are used to develop this spreadsheet for the first 17 years:

Year	Interest	Amount
1	$7.00	$107.00
2	$7.49	$114.49
3	$8.01	$122.50
4	$8.58	$131.08
5	$9.18	$140.26
6	$9.82	$150.07
7	$10.51	$160.58
8	$11.24	$171.82
9	$12.03	$183.85
10	$12.87	$196.72
11	$13.77	$210.49
12	$14.73	$225.22
13	$15.77	$240.98
14	$16.87	$257.85
15	$18.05	$275.90
16	$19.31	$295.22
17	$20.67	$315.88

10. After 10 years, there will be $196.72 in the account.

11. The principal will triple during the 17th year.

Choose a Calculation Method, pages 84–85

Calculation methods and reasons may vary. Possible responses are given.

1a. mental math

1b. An exact answer is needed quickly, and it would be quite inconvenient to use a calculator or paper and pencil.

1c. $13.25

2a. calculator

2b. Many exact answers from tedious computations are needed. A calculator should be readily available.

2c. $110.49

3a. estimation

3b. While planning purchases in a store, it may be good enough to estimate their total cost. It's quite possible that a calculator or paper and pencil would not be readily available.

3c. No. An estimate might reveal that the total cost is about $13 + $6 + $2, or $21.

4a. estimation

4b. The exact amount of time it should take you to get to Joey Town does not seem to be critical, so an estimate might be good enough.

4c. About $1\frac{1}{2}$ hours. The car travels 55 miles in 1 hour. The additional 33 miles is about one half of 55, so it should take $(1 + \frac{1}{2})$ hours.

5a. paper and pencil or a calculator

5b. An exact answer is usually needed when dealing with recipes. Paper and pencil is probably the best method—unless you have a calculator that handles fractions. If a permanent record of the results is needed (for future use), then paper and pencil should probably be used.

5c. $1\frac{5}{16}$ cups of flour; $\frac{3}{8}$ cup of nuts

6a. estimation

6b. An exact answer for a tip is not generally needed.

6c. about $2. Think of 15% as 10% + 5%. First mentally approximate 10% of the total: 10% of $12.76 is about $1.30. Since 5% is half of 10%, 5% of $12.76 is about $0.65. So 15% is about $1.30 + $0.65, or about $2. (It might be more convenient to leave a tip of $2 than, say, $1.95.)

7a. paper and pencil

7b. You might want to draw a diagram to help you make a list of all the games in the tournament.

7c. 6 games. The following diagram for players A, B, C, and D can be used to show the games in the tournament:

A—B A—C A—D
B—C B—D C—D

What's Inside a Rectangle?, pages 86–87

1. 96 ft²
2. 1 ft²
3. Sample: a 6-by-4 and an 8-by-3 rectangle
4. Sample: a 5-by-4 and an 8-by-1 rectangle
5. 70 m
6. 7.5 acres
7a. 100 m²
7b. 10,000 m²
8a. 10,000 m²
8b. 100 ares
8c. 10,000 ares
9. 192 cm²
10. 36 yd²

Base Times Height—and You'll Be Right!, pages 88–89

1. 120 ft²
2. 3 ft²
3. $4\frac{1}{2}$ ft²
4. $7\frac{1}{2}$ ft²
5. $8\frac{1}{4}$ ft²
6a. 0.025
6b. 0.075
6c. 0.706
7a. rectangle
7b. 4.32 m²
8. $2\frac{1}{2}$ ft
9. 20 yd
10. 20.54 m²
11. $10\frac{3}{8}$ in²
12. 25 m²

Going Around in Circles, pages 90–91

1. 75.36 in.
2a. 81.64 in.
2b. 80.07 in.
2c. worse
2d. The car will travel less distance for the same number of revolutions.
3. 31,400 mi²
4a. Estimates will vary, but should be between 76 and 80 square units.
4b. 78.5 square units
4c. 21.5 square units
5a. one 14-in. pizza
5b. 76.93 in² more pizza
6. 50.24 m

Estimates for Questions 7–8 may vary.

7. about 12 square units
8. about 6 units
9. 40.84 m
10. 37.74 m²

Turn Up the Volume, page 92–93

1a. 49
1b. 50 cubic units
1c. 27 cubic units
2. No. Gina dug 125 ft³ of dirt. Tina dug 2 x 2.5 x 2.5 x 2.5, or 31.25 ft³ of dirt.
3a. 2964 cm³; 4872 cm³
3b. The 15-ounce box contains more cereal per cubic centimeter of space. (For each box, divide the weight of the cereal by the volume of the box. In the 15-ounce box, there is about 0.005 ounce of cereal per cubic centimeter of space. In the 18-ounce box, there is about 0.004 ounce of cereal per cubic centimeter of space.)
4. 83,116,800 ft³
5. 3 times as great
6. 1177.5 cm³
7. 588.75 cm³
8. 2355 cm³
9. Yes
10. No. Doubling the radius (for a fixed height) gives you 4 *times* the volume. (Suppose *r* is the radius. Doubling the radius yields 2*r* for the new radius. When this radius is squared, the result is (2*r*) x (2*r*) = 4*r²*. This is 4 times the original radius squared.
11. 3 times as great
12. 90 cm³
13. 87.92 cm³
14. 84.9 cm³
15. the cone (The cone gives about 80 cm³ of ice cream per $1, versus about 67 cm³ from the bar and about 77 cm³ from the cylinder.)

From *Math for Real Kids* © 2005 David B. Spangler.

Estimation with Area and Volume, pages 94–95

All estimates in this lesson may vary.

1. An estimate might be somewhere from 110 to 130 bushels per acre.
2. about 3 liters
3. about 700 pounds
4. about 21 square units
5. 2 to 3 to 4
6a. about 1,000 beans
6b. Sample: The length, width, and height are each about "10 beans long." 10x10x10 = 1,000.
7. 1107 mm², or 1.107 m²
8a. about 16,800 ft³
8b. Sample: Suppose you assume that the average depth of the pool is 4 feet. Then the pool could be viewed as a rectangular prism with length 75 feet, width 56 feet, and depth 4 feet. If rounded values are used in the computation, the estimated volume might be 70 x 60 x 4, or about 16,800 ft³. (Since 56 is rounded up to 60, rounding 75 down to 70 compensates for some of the "overage.") Another Sample: Suppose you assume that the depth of the pool gradually gets deeper as you go from the shallow end (3 feet) to the deep end (5 feet). Then the pool could be viewed as a trapezoidal prism. The trapezoid (side view of the pool) would have bases of 3 feet and 5 feet, with a height of 75 feet. Thus, the area of the trapezoid would $\frac{(3+5)}{2}$ x 75, or 300 ft². The volume of the trapezoidal prism would be 300 x 56, or about 180,000 ft³ (if 56 is rounded to 60).
8c. About 125,680 gallons of water. If 7.481 is rounded down to 7 and 16,800 is rounded up to 20,000, you would have 7 x 20,000, or about 140,000 gallons of water.
8d. About 1,043,144 pounds. If 8.3 is rounded down to 8 and 140,000 gallons is used from 8b, the result is about 1,120,000 pounds.

Hidden Figures, pages 96–97

1. Number of squares that are a 1-by-1 *Large Square*: 16
   Number of squares that are a 2-by-2 *Large Square*: 9
   Number of squares that are a 3-by-3 *Large Square*: 4
   Number of squares that are a 4-by-4 *Large Square*: 1
   Number of squares that are a 1-by-1 *Small Square*: 24
   Number of squares that are a 2-by-2 *Small Square*, and are completely bounded by a heavy border: 5
   Number of squares that are a 2-by-2 *Small Square*, where the square has exactly two sides bounded by a heavy border: 4
   Other: Number of squares that are a 2-by-2 *Small Square*, and have no sides bounded by a heavy border: 4

   Total Number of Squares 67

2. 47
   Triangles may be listed in any order. In the list below, all triangles with a vertex at *A* are listed first, followed by all those with a vertex at *B* (and no vertex at *A)*, and so on.
   △ABG, △ACH, △ACD, △ABH, △ABE, △AGF, △AHF, △AHE, △ADE, △AFC, △ABD, △ADF, △ABF, △ACE.
   △BHC, △BHD, △BED, △BFD, △BHF, △BFC, △BDJ, △BCI, △BIH, △BFI, △BFJ, △BGH, △BEF, △BGD, △BCD, △BCE.
   △CID, △CDH, △CEH, △CFE, △CDF
   △DGF, △DHF, △DFE, △DHE, △DIH, △DJE, △DJH, △DFI.
   △EHF, △EJF.
   △FHG, △FJH.

Tackling Integers 1, page 98

1. 2
2. -2
3. -10
4. -1
5. 4
6. 0
7. 4
8. -9
9. -10
10. gained 4 yards (+4 yards)
11. lost 10 yards (-10 yards)

Tackling Integers II, page 99

1. -6
2. 2
3. -4
4. 0
5. -8
6. 11
7. -5
8. 9
9. -11
10. gained 4 yards (+4 yards)
11. lost 19 yards (-19 yards)

Wind Chill Numb-brrs, page 100

1. 20 miles per hour
2a. 34 degrees colder
2b. 28 degrees colder
2c. 19 degrees colder
3a. 43 degrees colder
3b. 35 degrees c older
3c. 24 degrees colder
4. Estimates may vary. A reasonable estimate is -15°F
because in the table a wind chill temperature of -60°F falls halfway between the actual
temperatures of -10°F and -20°F.
5. -40°F
6. 16°F

Positively Negative, page 101

1. Total Scores: Your School, 177; School X, -12; School Y, -96
2. 189 points
3. 84 points
4a. -24 pounds
4b. -2 pounds per month
5. -$20
6a. -89.2°C
6b. -235°C
6c. -273.1°C

Getting to the Root of the Problem, page 102

1a. 1.73
1b. 2.65
1c. 3.32
1d. 5.48
1e. 8.06
1f. 11.96
2a. 38.58 miles
2b. 86.27 miles
2c. 211.32 miles

Pythagorean Problems, pages 103–104

1a. 16 square units
1b. 9 square units
1c. 25 square units
2. The sum of the areas of the squares on sides $\overline{AB}$ and $\overline{BC}$ is equal to the area of the square on side $\overline{CA}$.
3a. 50-inch TV
3b. 46-inch TV
3c. 27-inch TV
4. 41.2 feet
5. 75 miles
6. 75 feet
7. 6.7 units

From *Math for Real Kids* © 2005 David B. Spangler.

Number Tricks, pages 105–106

Computations in the Arithmetic columns will be based on the first numbers picked.

Number Trick 1

Verbal	Algebra	Arithmetic
a. Pick a number.	x	**Let's pick 7.**
b. Add 5.	$x + 5$	$7 + 5 = 12$
c. Multiply by 2.	$2(x + 5) = 2x + 10$	$2 \times 12 = 24$
d. Subtract 10.	$2x + 10 - 10 = 2x$	$24 - 10 = 14$
e. Generalization: **You will end up with twice the number that was picked.**		

Number Trick 2

Verbal	Algebra	Arithmetic
a. Pick a number.	x	**Let's pick 9.**
b. Multiply by 4.	$4x$	$4 \times 9 = 36$
c. Subtract the number you picked.	$4x - x = 3x$	$36 - 9 = 27$
d. Add 12.	$3x + 12$	$27 + 12 = 39$
e. Multiply by $\frac{1}{3}$.	$\frac{1}{3}(3x + 12) = x + 4$	$\frac{1}{3} \times 39 = 13$
f. Generalization: **You will always end up with 4 more than the number that was picked.**		
g. Describe how you could try out Number Trick 2. **Sample answer: Have someone do Steps a–e mentally. When the person tells you his or her final result, subtract 4 to reveal the number the person picked.**		

Number Trick 3

Verbal	Algebra	Arithmetic
a. Pick a number.	x	**Let's pick -3.**
b. Add the number you picked.	$x + x = 2x$	$-3 + -3 = -6$
c. Subtract 2.	$2x - 2$	$-6 - 2 = -8$
d. Multiply by 2.	$2(2x - 2) = 4x - 4$	$2(-8) = -16$
e. Multiply by $\frac{1}{4}$.	$\frac{1}{4}(4x - 4) = x - 1$	$\frac{1}{4}(-16) = -4$
f. Subtract the number you picked.	$x - 1 - x = -1$	$-4 - -3 = -1$
g. Generalization: **You will always end up with -1.**		

Planely Algebra, pages 107–108

1–2. The line of fit may vary depending upon how it is estimated.

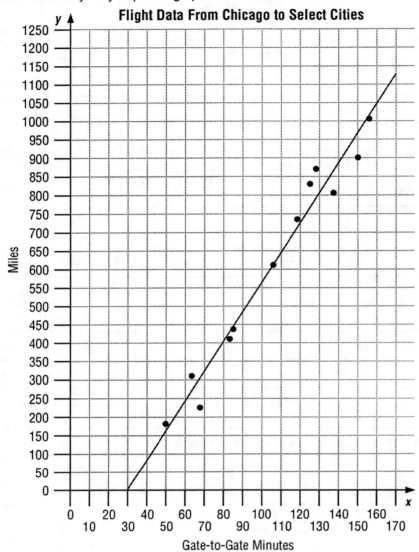

Flight Data From Chicago to Select Cities

3a. 1,125 miles   **4a.** 80 miles   **5a.** (30,0)
3b. 90 minutes   **4b.** 8 miles per minute   **5b.** taxiing on the runways

Upside-Down Calculator Problems,
pages 109–110

1. hOg·S   4. ·SIgh   6.
2. ELIgIBL·E   5. ·SELL
3. BO·ISE

1.g	O	2.B	B	L	3.E
I	■	L	■	■	L
g	■	I	■	■	I
4.g	O	g	h	■	S
L	■	5.h	O	S	E
E	■	■	■	■	■

Printed in the United States
91162LV00006B/24/A